CONTENTS

Dr Una Coales's MRCGP AKT Hot Topics

Second Edition

Una Coales BA (Hon) MD FRCSEd FRCSEd (ENT) FRCGP DRCOG DFFP PGCertMedEd

www.lulu.com

First edition published in Great Britain by www.lulu.com 2010.
Second edition published 2011.

Published and distributed by Lulu.com
3101 Hillsborough Street,
Raleigh, NC 27607, USA.
Website: www.lulu.com

Preface

This is the second edition of the perfect study guide for the MRCGP AKT exam and allows you to focus your revision on topics relevant for the AKT exam. I have published 11 medical exam revision books and personally taught the MRCGP exam to over 3000 International and UK established and training GP trainees. Not only do you need Entwistle's deep knowledge to score highly but also Entwistle's strategic knowledge. The number one reason for failing AKT is lack of knowledge of statistics. The number two reason is weak clinical knowledge. By spending 150 to 200 hours digesting the 'strategic' contents of this book and expanding your 'deep' knowledge with the relevant sources cited, as necessary, you will pass your AKT first-time.

For those who wish to be personally tutored through the AKT exam, please contact me on www.mrcgpcourses.co.uk.

Good luck! As I may only accommodate a finite number of candidates on my private courses, so this book aims to reach all 3000 UK GP specialist trainees and international GPs preparing for the AKT exam for MRCGP (UK) or MRCGP (INT). Read the study plans of top-scoring AKT course candidates. Allocate 150 hours to hard work, determination and dedication and acquire the knowledge of the AKT!

For those preparing for CSA, please read the Dr Una Coales's MRCGP CSA book available through lulu.com.

Dr Una Coales BA (Hon) MD FRCSEd FRCSEd (ENT)
FRCGP DRCOG DFFP PGCertMedEd
Portfolio NHS GP and Private Educator and Author
www.mrcgpcourses.co.uk
mrcgpcourses@aol.com

November 2011

Study plans from top-scoring AKT Course GP STs

'Dear Una, I've set out below what I did to revise, in case it is helpful. I think I ended up using my background knowledge as in the last three years I've done MRCP, DCH, DRCOG and attended the theory course for the DFFP. My VTS scheme has got some good posts so I've also done ENT, ophthalmology, urology, orthopaedics, dermatology and rheumatology jobs as well as the standard paeds and O and G. (I did 18 months medicine and 6 months A and E before starting the GP scheme). So the main holes in my knowledge are psych and actual GP as I am in ST2.

Anyway, this is what I did to revise:
Your notes, course questions and book (of course!)
Passmedicine.com - free site with good explanations (I wonder who runs it for free?)
nPEP - free with AIT number and of similar style to the exam

Question books:
nMRCGP Practice Papers: AKT by Daniels- questions are harder than the exam but covered some of the admin/management type questions well
nMRCGP AKT Study Guide by Khan, Jabbour and Rehman
AKT for the new MRCGP by A-Ali for examples of algorithm questions although there weren't actually any algorithms in the exam!

Other questions: AKT Questions from the RCGP website (discussed as a group with people on my VTS scheme); Onexamination.com - lots of questions but not very similar to exam (compared with for MRCP and DRCOG where a number of the questions on this site were repeated word for word). I tried to read around the question if I didn't know the answer.

I also did some background reading:
- The topics mentioned by the RCGP in their feedback from previous AKT exams
- NICE clinical guidelines (just the summaries) and main SIGN guidelines but I think only asthma and COPD came up in the exam
- OHGP and some of the Oxford General Practice Library (OGPL) books which basically have the same information as OHGP but in a bit more depth and in colour (which helps to keep you awake).
I thought that the Women's Health book was good as I didn't have enough time to reread the books
I had read for DRCOG
- BNF guidance on prescribing (1st section) - definitely useful
- Essential Revision Notes for MRCP - Kalra. I flicked through this as I

had read it for MRCP.
- For stats I read Harris and Taylor which you recommended, parts of Stacey and Toun- Critical Reading Questions for the MRCGP and Essential Statistics for Medical Examinations by Faragher. None of these were necessary as I could do all the questions from your course material. I'd advise people not to waste time getting bogged down in Faragher because the questions were much simpler than that. Hope this is useful. Best wishes, RW MRCP.' July 2008
RW, ST2, Eastern Deanery, 94.95% AKT May 2008,
TOP AKT SCORE (188/198) in UK out of 1163.
Achieved **12/12 CSA passes** in 2009 and became a **partner** Jan 2010.

'Dear Una,
I started about 8 weeks before the exam. I worked mainly at weekends and took a week study leave just before the exam so I could cram a bit more in! Below is a list of things I used to revise.

Did questions on passmedicine twice. Read relevant NICE guidelines-guided quite a lot by passmedicine -although these came up in the exam a lot less than I thought they would
Read Una Coales manual through once and then some sections for a second time
Did questions from Una Coales course
Read parts of Oxford handbook of GP focusing on topics I felt weak on-eg ENT, ophthalmology, dermatology, infectious diseases, admin
Contraception section in BNF
DVLA and fitness to fly guidelines
Did RCGP essential knowledge updates (on RCGP website). Free questions from AiT (150 questions)
Lots of questions from books-Nuhzet A-Ali, Get Through MRCGP: AKT by Dianne Campbell, NMRCGP Applied Knowledge Test Study Guide: Sample Questions and Explanatory Answers (Masterpass) by Aalia Khan, Ramsey Jabbour, and Almas Rehman
First 20 or so pages of the BNF. Medical Statistics made Easy-by Harris and Taylor
Things I didn't read but would be useful as came up in the exam: Good medical Practice guide from GMC-a few questions mentioned this.
Oxford handbook of medicine (maybe even more useful than GP one?)
- exam v medical-needed to draw on a lot of old med school knowledge.
Read clinical chemistry section-quite a few questions interpreting blood results eg calcium etc.
Overall, the exam was quite different from what I'd expected. It did not focus on the NICE guidelines as much as I thought it would and a lot of it was clinical problem solving.

Anyway, I hope this is helpful. Thank you again for your excellent course and handbook. It really helped me to focus my revision.'
Dr NT, ST3 London Deanery, 92.9%, AKT Oct 2009.

'Hi Una, Hope you're well. I attended your AKT course in December. Sorry for the delay, just to let you know that I passed and I also managed to get the top score in this sitting (92.5%) so thank you very much for your help.

As a word of advice to other candidates I used only the following: your course material, Passmedicine, the free questions from the RCGP, selected quick reference NICE guidelines and dipped into the OHCM occasionally. I think if you know and understand the course material and the explanations on Passmedicine you would pass.

Anyhow, thanks again. Best Wishes N.'
Dr NG, ST2 London Deanery
(92.5%, the highest score out of 915 STs, Jan 2010)

'Hi Una,
I have included some details of how I revised in case it is helpful to other people. I am not much of a crammer and probably used some of the knowledge which I have acquired from my 2 yrs of GP training so far. I have been very fortunate and done a wide variety of jobs. In the two yrs (I am ST2) I have completed my DRCOG, DCH and DFSRH so I think that the time that I spent studying for these probably helped. Since ST1 I have always read the InnovAiT magazine which covers many curriculum areas well. I have also completed the essential knowledge updates from the college as they are produced. I probably started revising about 12 weeks before the exam. Below is a list of the resources I used:

Online resources:
Passmedicine site-completed all the questions once and repeated the ones I had got wrong
Onexamination questions-I don't think these are as good as passmedicine
nPEP questions-free with AiT number-300 Qs
InnovAiT-free online questions can be quite difficult to navigate to these on their website but a really valuable resource and free
DVLA guidelines-read the online guidance
CAA guidelines-read the online guidance
NICE quidelines-read the summaries for all the clinical guidelines
SIGN guidelines-read the asthma guideline
RCGP website-completed the 50 example Q's-however there are no

answers to these on the website

Books
Una Coales handbook-I read through this twice before the exam
BNF-read 1st 20 pages on prescribing guidance
Oxford Handbook of GP-read through this once
Medical stats made easy (M.Harris and G.Taylor)-very easy book to read although I don't think this is essential as Una's course covers all the stats you need to know
ENT in Primary Care (Peter Robb, Alex Walson) (my weakest area)-makes ENT seem really easy-is useful for general everday ENT knowledge
nMRCGP Practice Questions: AKT (Rob Daniels)-the Q's in here weren't as good as online resources and weren't as up to date

Hope this is useful. Kind Regards,'
CB, ST2 KSS Deanery (90%, AKT April 2010)

'Dear Una, I have not done the MRCP fortunately! I have come through the foundation system to VTS. I have a 1 and a half yr old at home so just aimed to do 1-2 hrs a day on week days, a little more a weekends, and took 4 days study leave before the exam. I did all the pass medicine questions twice, nearly 3 times and used the "get through the AKT" and another question book. I found the stats on your course extremely helpful especially the tricks to do the calculations quickly and interpreting odds ratios as a percentage. I also read the med stats made easy book and some nice guidelines and used CK, S as well as your handbook. I think in total I probably did about the 150hrs you recommended.
Thanks for your help H.'
Dr HW, ST3 London Deanery (92.5% AKT April 2010)

ADULT BASIC LIFE SUPPORT (2010)
Mnemonic: DRSABCC

Ensure Safety of Self & Patient/ Assess for **D**anger

↓

Check for **R**esponse – shake and shout 'are you okay?'

↓

Shout for help 'someone I need some help here' – open pt's **A**irway: head tilt, chin lift or jaw thrust

↓

If breathing, place in recovery position←—Check **B**reathing: Look, listen & feel for 10 seconds.

↓

If not breathing, **C**all '999' (or '911') on your mobile.

↓

Give 30 **C**hest compressions

↓

Give 2 rescue breaths and continue **CPR at ratio of 30:2** at 100 compressions a minute.
Continue until: help arrives, patient shows signs of life, or you become exhausted.

ANAPHYLAXIS ALGORITHM (Resuscitation Council UK)
Anaphylactic reaction?

↓

Airway, Breathing, Circulation, Disability, Exposure

↓

Dx – look for: acute onset of illness, life-threatening airway and/or breathing and/or circulation problems (1) and usually skin changes.

↓

Call for help, lie patient flat, raise patient's legs

↓

Adrenaline (2)

↓

When skills and equipment available:
Establish airway, high flow oxygen, IV fluid challenge (3),
chlorphenamine (4), hydrocortisone (5)
Monitor: pulse oximetry, ECG, BP

(1)   Life-threatening problems: Airway: swelling, hoarseness, stridor.
Breathing: rapid, wheeze, fatigue, cyanosis, SpO_2 < 92%, confusion.
Circulation: pale, clammy, low BP, faintness, drowsy/coma.
Adrenaline IM unless experienced with IV adrenaline. IM doses of
1:1000 into anterolateral thigh or upper arm and may be repeated at 5

min intervals according to BP, P and RR. Subcutaneous injection is NOT recommended. 1/2 dose if on TCAs or MAOIs (phenelzine).

	(2) Resuscitation Council	(2) BNF
Adult	500 mcgs IM (0.5ml)	same
Child > 12 yo	500 mcgs IM (0.5ml)	same
Child 6-12 yo	300 mcgs IM (0.3ml)	6-12 yo 250 mcgs IM (0.25 ml)
Child < 6 yo	150 mcgs IM (0.15 ml)	6 mos-6 yo 120 mcgs IM (0.12 ml)
		< 6 mos 50 mcgs IM (0.05 ml)

Adrenaline IV to be given only by experienced specialists.
Titrate: Adults 50 mcgs; Children 1mcg/kg

(3) IV fluid challenge: Adult 500-1000 mL; Child crystalloid 20 mL/kg. Stop IV colloid if this might be the cause of anaphylaxis.

	(4) Chlorphenamine (IM or slow IV)	(5) Hydrocortisone
Adult or child > 12 yo	10 mg	200 mg
Child 6-12 yo	5 mg	100 mg
Child 6 months to 6 yo	2.5 mg	50 mg
Child < 6 months	250 mcgs/kg	25 mg

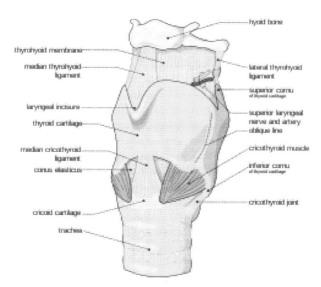

Cricothyroidotomy: Insert an emergency airway between the cricoid and thryroid cartilages.

ALZHEIMER'S DIS (NICE Nov 06)

- **Donepezil, galantamine, rivastigmine for mild to moderately severe Alzheimer's disease (AD).**
- **Memantine for moderately severe to severe AD, only as part of well designed clinical studies.**

Patients with mild AD who are currently receiving donepezil, galantamine or rivastigmine, and patients with moderately severe to severe AD currently receiving memantine, whether as routine tx or as part of a clinical trial, may continue on tx (including after the clinical trial) until they, their carers +/or specialist consider stopping.

The benefits of these drugs for vascular dementia or dementia with Lewy bodies have not been assessed here.

The 3 acetylcholinesterase inhibitors donepezil, galantamine and rivastigmine are recommended as options in the mx of AD of mod severity only (that is, those with a [MMSE] score of 10-20 points) and the following:

Only psychiatrists including those specialising in learning disability, neurologists, and doctors specialising in the care of the elderly) should initiate tx. Seek carers' views on patient's baseline condition.

Patients who continue on the drug should be reviewed q 6 months by MMSE score and global, functional and behavioural assessment. Seek carers' views on the pt's condition at FU. The drug should only be **continued while the pt's MMSE score remains ≥ 10 points** and their global, functional and behavioural condition remains at a level where the drug is considered to be having a worthwhile effect. Any review involving MMSE assessment should be undertaken by an appropriate specialist team, unless there are locally agreed protocols for shared care. When the decision has been made to prescribe an acetylcholinesterase inhibitor, it is recommended that tx should be initiated with a drug with the lowest acquisition cost. However, an alternative acetylcholinesterase inhibitor could be prescribed where it is considered appropriate having regard to adverse event profile, expectations around concordance, medical comorbidity, possibility of drug interactions and dosing profiles.

Clinical need and practice

Dementia is a chronic progressive mental d/o that adversely affects higher cortical functions including memory, thinking and orientation. AD is the most common form of dementia. It is a degenerative cerebral disease with characteristic neuropathological and neurochemical features. AD is usually insidious in onset and develops slowly but steadily over a period of several yrs. Progression is characterised by deterioration in cognition (thinking, conceiving, reasoning) and functional ability (ADL) and a disturbance in behaviour and mood. Changes in ≥ 1 of these domains and their effects on the pt provide the basis for dx and they are used to assess the severity and progression of the condition. Evidence suggests that AD progression is dependent on age, and the time from dx to death is about 5–20 yrs (median 5 years in patients aged 75–80 yo).

AD lose the ability to carry out routine daily activities - dressing, toileting, travelling, handling money and many require a high level of care. Often, this is provided by an elderly relative, whose own health and quality of life is affected by the burden of providing care. Behavioural changes (aggression) are particularly disturbing for carers.

Non-cognitive symptoms in dementia include agitation, behavioural disturbances (i.e., wandering or aggression), depression, delusions and hallucinations.

Several different methods are used to assess the severity of AD. These include: the Clinician's Interview-based Impression of Change (CIBIC) and CIBIC-plus for global outcomes; the Progressive Deterioration Scale (PDS) for functional/quality-of-life scales; and the AD Assessment Scale – cognitive subscale (ADAS-cog – 70 points) or the MMSE – 30 points) for cognitive outcomes. MMSE score denotes the severity of cognitive impairment as follows: **mild AD: MMSE 21 to 26**
> **moderate AD: MMSE 10 to 20**
> **moderately severe AD: MMSE 10 to 14**
> **severe AD: MMSE less than 10.**

Pop data (2002) for England and Wales show an estimated prevalence of 290,000 with AD. On the basis of these figures a PCT with a pop of 200,000 might expect to have 1100 cases of AD. The incidence rate for AD in patients > 65 years has been estimated at 4.9 per 1000 person-years in the UK. The incidence rate appears to have been stable over the past 2 decades and is found to be related to age (rising with ↑ age) and gender (F > M). Approx 50–64% have mild to moderately severe AD, and 50% have mod severe to severe AD.

Patients with mild dementia are sometimes able to cope without assistance, but as the disease progresses, all eventually require the aid of carers, and 1/2 need residential care. **The total cost of care for dementia is estimated by the Audit Commission to be £6 billion/ yr in England, with 1/2 attributed to health and social services.**

Patients with dementia usually present to their GP with memory problems, and an estimated 39% to specialist clinics. The **role of memory clinics (NSF for Older People)** states that referral to specialist MH services should be considered in a number of circumstances for those with suspected dementia, not only for consideration of tx but also, i.e., if the dx is uncertain, if certain behavioural and psychological sxs are present, or if there are safety concerns with anti-dementia drugs, in accordance with local protocols.

Acetylcholinesterase inhibitors: donepezil, galantamine, rivastigmine: ↑ the concentration of ACh at sites of neurotransmission. **Donepezil (Aricept) is a specific and reversible inhibitor of AChE,** licensed in the UK at a dosage of 5 mg/day and 10 mg/day for the symptomatic tx of people with mild to moderately severe AD. In 2003, 77% of rx's for AChE inhibitors were for donepezil.

Galantamine (Reminyl) is a selective, competitive and reversible inhibitor of AChE, licensed in the UK for the symptomatic tx of mild to mod severe dementia of AD. In addition, galantamine enhances the intrinsic action of acetylcholine on nicotinic receptors, probably through binding to an allosteric site of the receptor. The maintenance dosage is 16–24 mg od. **Rivastigmine (Exelon) is an acetylcholinesterase and butyrylcholinesterase inhibitor, licensed in the UK for symptomatic tx of mild to mod severe AD.** The usual maintenance dosage is 3–6 mg bd. Typical side-effects of donepezil, galantamine and rivastigmine are related to the GI tract (i.e. N/V). These side-effects are dose related and although they are usually s-t they can lead to non-adherence

Memantine (Ebixa) is a voltage-dependent, moderate-affinity, uncompetitive N-methyl-D-aspartate (NMDA)-receptor antagonist that blocks the effects of pathologically elevated tonic levels of glutamate that may lead to neuronal dysfunction. It is used in the tx of mod to severe AD. The maintenance dosage is 10 mg bd. In clinical trials in mild to severe dementia, involving patients on memantine and patients on placebo, the most frequently occurring adverse events of mild to mod in severity with a higher incidence on memantine vs. placebo were dizziness, HA, constipation and somnolence.

The Committee concluded that the **prescribing of AChE inhibitors for people with AD and moderate cognitive impairment (MMSE scores between 10 and 20) is cost effective.** The Committee concluded that the evidence to support the assertion that memantine prevents patients with moderately severe to severe AD from being institutionalised was currently insufficient. On the basis of current evidence on clinical effectiveness **memantine could NOT reasonably be considered a cost-effective therapy for moderately severe to severe AD.**

ANALGAESIA

Breast pain	Gabapentin (DA agonist) down-regulates prolactin
Bone pain from mets	NSAIDs for breakthrough pain for CA and for pain of Paget's disease
Dysmenorrhoea (fibroids)	mefenamic acid
Mechanical back pain	paracetamol
Migraine	sumatriptan (5HT1 agonist) if simple analgaesia fails; not ergot alkaloids
Postherpetic neuralgia	amitriptyline (1st-line); gabapentin (2nd-line); no neuralgia: paracetamol
Raynaud's phenomenon	nifedipine (to ↓vasospasm and attacks)
Renal colic	pethidine or diclofenac IM
Sickle cell crisis	morphine; not pethidine, which can precipitate fits
Trigeminal neuralgia	carbamazepine

ANTIBIOTICS

Amoxicillin	acute bronchitis, exacerbation of COPD, OM; bacteriocidal; pneumonia; sinusitis; cellulitis
Benzylpen IV	suspect N meningitidis 1200 mg; 600 mg (1-9 yo); give cefotaxime (3rd gen) if allergic
Ciprofloxacin	acute pyelonephritis or use co-amox; gonorrhoea-treat with cipro 500 mg STAT or if pregnant, give amoxycillin 3g + probenecid 1g oral STAT, test for cure; PID (cipro 500 mg STAT + doxycycline 100 mg bd for 2/52 + metronidazole 400 mg bd for 5/7)
Co-amoxyclavulanic acid	human bites (URT aerobes + anaerobes); animal bites (pasteurella multicoda, capnocytophagia canimorsus, anaerobes, call CCDC for rabies, check tetanus)
Doxycycline	chlamydia trachomatis 100 mg bd x 1/52 or azithromycin 1g STAT; test for cure 1mo
Erythromycin	atypical pneumonia; bacteriostatic; clarithromycin to non-responders, rx acne in pregnancy
Flucloxacillin	folliculitis (Staph aureus + Gram negative org); impetigo (Staph aureus, strep pyogenes)
Metronidazole	BV (gardnerella – Gram negative anaerobe); acigel for recurrent cases; trich vaginalis 400 mg

	bd for 5/7; triple tx; antabuse reaction; tx antibiotic-associated clostridium difficile
Minocin MR	100 mg acne; **Oxytetracycline** 500 mg bd; 2nd line or pregnant: erythromycin 500 mg bd

Phenoxymethylpenicillin + flucloxacillin for cellulitis (staph aureus, strep pyogenes); acute tonsillitis (pen)

Trimethoprim	uncomplicated UTI 200 mg bd for 3/7; alternative nitrofurantoin 50 mg 6hly; if pregnant treat with cefadroxil 500 mg bd
Antimalarials	Start Malarone 1-2 days prior to trip and continue for 1 week post return. Mefloquine start 2-3 weeks prior to trip and continue for 4 weeks post return. Usual prophylaxis starts 1 week prior and continues for 4 weeks post return. Can have disease up to 3/12 post.

Podophyllotoxin or imiqimod cream genital warts

ANTENATAL CARE (NICE Mar 2008)

Folic acid 5 mg if has coeliac or other malabsorptive state, diabetes, epilepsy, sickle cell anaemia, thalassaemia.

Lifestyle considerations *New* All women should be informed at the booking appt about the importance for their own and their baby's health of maintaining adequate vitamin D stores during pregnancy and whilst breastfeeding. Women may choose to take 10 mcgs of vitamin D per day, as found in the Healthy Start multivitamin supplement. Enquire as to whether women at greatest risk are following advice to take this daily supplement. These include: South Asian, African, Caribbean or Middle Eastern family origin; limited exposure to sunlight, predominantly housebound, or usually remain covered when outdoors eat a diet particularly low in vitamin D, who consume no oily fish, eggs, meat, vitamin D-fortified margarine or breakfast cereal; a pre-pregnancy BMI > 30 kg/m^2.

Screening for haematological conditions

New Screening for sickle cell diseases and thalassaemias should be offered to all ASAP in pregnancy (ideally by 10 weeks). The type of screening depends upon the prevalence and can be carried out in either 1°or 2° care.

Screening for foetal anomalies *New* Participation in regional congenital anomaly registers and/or UK National Screening Committee-approved audit systems is strongly recommended to facilitate the audit of detection rates.

New **The 'combined test'** (nuchal translucency, beta-human chorionic gonadotrophin, pregnancy-associated plasma protein-A) should be offered to **screen for Down's syndrome** between 11 wks and 13 wks 6 days. For women who book later in pregnancy the most clinically and cost-effective serum screening test (triple or quadruple test) should be offered between 15 wks and 20 wks.

Screening for clinical conditions

New Screening for gestational diabetes using RFs is recommended in a healthy pop. At the booking appt, the following RFs for gestational diabetes should be determined and offered testing: BMI>30 kg/m^2; previous macrosomic baby weighing $\geq$ 4.5 kg; previous gestational diabetes (refer to 'Diabetes in pregnancy' [NICE clinical guideline 63]; fhx of diabetes (1st-degree relative with diabetes); family origin with a high prevalence of diabetes: South Asian (specifically women whose country of family origin is India, Pakistan or Bangladesh); black Caribbean; Middle Eastern (specifically women whose country of family origin is Saudi Arabia, United Arab Emirates, Iraq, Jordan, Syria, Oman, Qatar, Kuwait, Lebanon or Egypt).

First contact with a healthcare professional

Give info (supported by written info and antenatal classes) on: folic acid supplementation; food hygiene, including how to $\downarrow$ the risk of a food-acquired infection; lifestyle advice (smoking cessation, recreational drug use and EtOH consumption); all antenatal screening, including risks and benefits of the screening tests.

Booking appointment (ideally by 10 weeks)

Give the following info on: how the baby develops during pregnancy; nutrition and diet, including vitamin D supplementation; exercise, including pelvic floor exercises; antenatal screening, including risks and benefits of the screening tests; pregnancy care pathway ; place of birth (refer to 'Intrapartum care' [NICE CG 55]); breastfeeding, including workshops; participant-led antenatal classes; maternity benefits.

- identify those need additional care and plan pattern of care for the pregnancy
- check blood group and rhesus D status (ideally before 10 weeks)
- offer screening for haemoglobinopathies, anaemia, red-cell alloantibodies, hepatitis B virus, HIV, rubella susceptibility and syphilis (ideally before 10 weeks)
- offer urine tests to screen for asymptomatic bacteriuria and test for proteinuria
- inform pregnant women < 25 years about the high prevalence of chlamydia infection in their age group

- give details of their local National Chlamydia Screening Programme (www.chlamydiascreening.nhs.uk)
- offering screening for **Down's syndrome** ('combined test' at 11 weeks to 13 weeks 6 days serum screening test (triple or quadruple) at 15 weeks to 20 weeks).
- offer early ultrasound scan for gestational age assessment (crown–rump measurement 10 weeks to 13 weeks 6 days, head circumference if crown–rump length is > 84 mms) and for structural anomalies (normally between 18 weeks and 20 weeks 6 days).
- measure height, weight and calculate BMI; measure BP and test urine for proteinuria
- offer screening for gestational diabetes and pre-eclampsia using risk factors
- identify women who have had genital mutilation
- ask about any past or present severe mental illness or psychiatric tx
- ask about mood to identify possible depression
- ask about the woman's occupation to identify potential risks.

16 weeks: Review results of all screening tests; reassess planned pattern of care for the pregnancy and identify F who need additional care.

- Investigate Hb < 11 g/100 ml (consider iron), BP, test urine for proteinuria.
- Give info on routine anomaly scan.

18 to 20 weeks, if the F opts, scan for the detection of structural anomalies.

- If placenta is found to extend across the internal cervical os at this time, offer another scan at 32 wks.

25 weeks (nullips) measure and plot symphysis–fundal height; BP and test urine for proteinuria; give info.

28 weeks (all) give information

- offer a 2nd screening for anaemia and atypical red-cell alloantibodies
- investigate hb < 10.5 g/100 ml and consider iron
- offer anti-D prophylaxis to rhesus-negative women
- BP and test urine for proteinuria; measure and plot symphysis–fundal height

31 weeks (nullips) BP, test urine for proteinuria; measure and plot symphysis–fundal height; give info; review, discuss and record the results of screening tests undertaken at 28 weeks; reassess planned pattern of care for the pregnancy and identify women who need additional care.

34 weeks (all) Give info on: preparation for labour and birth, coping with pain in labour and the birth plan; recognition of active labour.

- offer a 2nd dose of anti-D to rhesus-negative women
- BP and test urine for proteinuria; measure and plot symphysis–fundal height
- review, discuss and record the results of screening tests undertaken at 28 wks; reassess planned pattern of care for the pregnancy and identify women who need additional care.

36-weeks (all) Topics to cover: breastfeeding info, including technique and good mx practices that would help a F succeed, i.e. detailed in the UNICEF 'Baby Friendly Initiative' (www.babyfriendly.org.uk)

- vitamin K prophylaxis and newborn screening tests
- care of the new baby; postnatal self-care; awareness of 'baby blues' and postnatal depression
- BP and test urine for proteinuria; measure and plot symphysis–fundal height
- check position of baby; babies in the breech presentation, offer external cephalic version (ECV)

38 weeks will allow for:

- BP and urine testing for proteinuria; measure and plot of symphysis–fundal height
- info giving, including options for mx of prolonged pregnancy

40 weeks (nullips):

- BP and test urine for proteinuria; measure and plot symphysis–fundal height
- give info about the options for prolonged pregnancy

For women who have not given birth by 41 weeks: offer info

- offer a membrane sweep
- offer induction of labour
- BP and urine tested for proteinuria; symphysis–fundal ht should be measured and plotted

General Throughout the entire antenatal period, be alert to risk factors, signs or sxs of conditions that may affect the health of the mother and baby, i.e. domestic violence, pre-eclampsia and 'Diabetes in pregnancy' [NICE 63].

RCOG guidance on Chickenpox in Pregnancy Test for VZV IgG if exposed to chickenpox and give VZIG w/n 10d of exposure if neg. UK Advisory Group on chickenpox advises give oral acyclovir if w/n 24h rash onset and > 20/40 if a pregnant F develops chickenpox. Fetal Varicella Syndrome (micropthalmia, cataracts, hypoplasia, skin scarring, neurology, mental retardation) can develop anytime between 3 and 28 weeks gestation. Prenatal dx of FVS may be performed at 16-20 weeks by ultrasound. If maternal infection occurs at term, delay elective delivery by 5-7 days. Check neonate for VZV IgM and at 7 months check VZV IgG. If delivery is before 5 days, give VZIG to neonate. Vaccinate after delivery in hospital but before discharge and again at 6 weeks postpartum.

ANTENATAL CARE (NICE Oct 2003)

Folic acid 400 mcg od x 12/40. Inform that vitamin A supplements are teratogenic.

Screening for Down's: 11-14 wks -nuchal translucency (NT), the combined test (NT, hCG and PAPP-A); 14-20 wks - triple test (hCG, AFP and uE3), the quadruple test (hCG, AFP, uE3, inhibin A); from 11-14 wks and 14-20 wks - the integrated test (NT, PAPP-A, +hCG, AFP, uE3, inhibin A), the serum integrated test (PAPP-A, +hCG, AFP, uE3, inhibin A)

Morning sickness - ginger, P6 acupressure, antihistamines; heartburn; constipation; haemorrhoids; varicose veins; 1/52 of topical imidazole for thrush; backache.

Routine use NOT recommended:

- Breast and pelvic exam. Antenatal EPDS (**Edinburgh postnatal depression** scale)
- Screen for toxo, hepatitis C, asymptomatic chlamydia, asymptomatic BV, GBS (group B strep), gestational diabetes
- Auscultation of the fetal heart. Formal fetal-movement counting U/S after 24 wks. Umbilical artery Doppler U/S for prediction of fetal growth restriction

10 weeks Advice on diet, lifestyle (no EtOH, tobacco, cannabis), long haul (DVT), maternity benefits, screening tests

- **Risk of listeriosis**: Drink pasteurised or UHT milk; avoid mould-ripened cheese (camembert, brie and blue-veined), pate, undercooked meals
- **Risk of salmonella**: avoid raw or undercooked eggs, mayo, raw meat especially poultry
- **Risk of toxo**: avoid cat faeces in litter or soil, wear gloves when gardening, cook raw meats, wash all fruits and vegetables, wash hands before preparing food
- **Identify women who may need additional care** (cardiac disease, HTN, renal disease, endocrine or IDDM, psych, haematological (thromboembolic or autoimmune antiphospholipid

syndrome), epilepsy on drugs, CA, severe asthma, drugs (heroin, crack, ecstasy), HIV or HBV, autoimmune, obesity BMI $\geq$ 35, > 40 yo and smoke, teens, recurrent miscarriages ($\geq$ 3 consecutive), preterm birth, severe pre-eclampsia, HELLP syndrome or eclampsia, rhesus isoimmunisation, uterine surgery (C/S, myomectomy, cone biopsy), antenatal or PPH on 2 occasions, retained placenta on 2 occasions, puerperal psychosis, grand multip (> 6 pregnancies), stillbirth or neonatal death, SGA infant (< 5%), LGA (> 95%), baby < 2500 g or > 4500 g, congenital anomaly.)

Screen for pre-eclampsia: nullip, > 40yo, FH, prior hx, BMI > 35, multiple pregnancies or pre-existing vascular disease (DM, HTN), sxs: HA, blurred vision, flashing before eyes, pain below ribs, sudden swelling of face, hands or feet]

ANXIETY (NICE Dec 2004)

Anxiety disorders are common, chronic, cause considerable distress and disability, often unrecognised and untreated.

Step 1: Recognition and diagnosis

- Apprehension, cued panic attacks, spontaneous panic attacks, irritability, poor sleep, avoidance, poor concentration.
- Panic d/o +/- agoraphobia - intermit episodes of panic or anxiety + taking avoiding action to prevent feelings
- Agoraphobia, social or simple phobia (not in guideline) episodes of anxiety triggered by ext stimuli
- Generalised anxiety d/o - over-arousal, irritability, poor concentration, poor sleep, worry about several areas most of time
- If a pt presents in A+E or other settings with a panic attack, ask if they are already receiving tx for panic disorder, undergo the minimum ix to r/o acute physical problems, not usually admit to a medical or psych bed, refer to 1° care for subsequent care, give written info about local, national voluntary and self-help groups.

Step 2: Treatment in primary care
Panic disorders

- Benzos are associated with a less good outcome in the long term and should NOT be prescribed for treatment.
- Sedating antihistamines or antipsychotics should NOT be prescribed.
- Interventions in order of longest duration od effect: CBT (7-14h total), SSRI licensed for panic disorder for at least 6 months after show improvement (or imipramine or clomipramine if no

improvement after 12 weeks or SSRI not suitable), self help (biblotherapy, review between every 4 and 8 weeks). Monitor pharmacological interventions: within 2 weeks of tx, 4, 6 and 12 weeks.

Generalised anxiety disorder

Immediate mx: Support and info, problem solving, benzos, sedating antihistamines, self-help.

Benzos should NOT be used beyond 2-4 weeks

Long term interventions: **CBT** (16-20h), SSRI (if no improvement after 12 wks, change to another SSRI), bibliotherapy based on CBT principles

Step 3: Review. Offer alternative tx at 12 weeks if no improvement

Step 4: Review and offer referral from primary care

Venlafaxine should only be initiated by a specialist or GPSI in MH for generalised anxiety. Dose should be no higher than 75 mg od. Before rx, check BP and undertake ECG. Now been revised and GPs may rx.

Step 5: specialist mental health services should conduct a thorough, holistic, reassessment of individual, environment and social circumstances

Monitoring: short, self-complete questionnaires (such as panic subscale of the agoraphobic mobility inventory for individuals with panic disorder) should be used to monitor outcomes wherever possible.

ACUTE ASTHMA MANAGEMENT (BNF 56, Sept 2008, p149)

Moderate acute asthma: able to talk, RR< 25/min (2-5yo ≤ 50/min, 5-12 yo ≤ 30/min). Pulse < 110bpm (2-5yo ≤ 130/min, 5-12 yo ≤ 120/min). $PaO_2 ≥ 92\%$. PFR > 50% predicted or best (5-12 yo ≥ 50% or best). Treat at home or in surgery and assess response to tx.

Tx: inhaled short-acting B2 agonist via large-volume spacer or O_2 driven nebulizer. Give 4- 10 puffs 100 mcg inhaled separately repeat 10-20 minutes or nebulised salbutamol 5 mg (2-5yo 2.5 mg, 5-12 yo 2.5-5 mg) or terbutaline 10 mg (2-5yo 5 mg, 5-12 yo 5-10 mg). Prednisolone 40-50 mg od x 5/7. Child 1-2 mg/kg. Monitor response for 15-30 minutes. If poor response or relapse < 3-4h, send to hospital.

Severe acute asthma: cannot complete sentences in 1 breath (child too breathless to talk or feed, use of access muscles), RR> 25/min (2-5yo > 50/min, 5-12 yo > 30/min). Pulse ≥110bpm (2-5yo >130/min, 5-12 yo >120/min). PaO_2 < 92%. PFR 33-50% predicted or best (5-12 yo < 50% or best). Send immediately to hospital.

Tx: High flow O_2. Inhaled short-acting B2 agonist via large-volume spacer or O_2 driven nebulizer (same as for moderate). Prednisolone 40-50 mg od or IV hydrocortisone 100 mg q 6h (< 1yo 25 mg, 1-5 yo 50 mg, 6-12 yo 100 mg). Monitor response for 15-30 minutes. If poor response, give ipratropium bromide via O_2 driven neb 500 mcgs (< 12 yo 250 mcgs). Consider IV B$_2$ agonist, aminophylline or magnesium sulfate. Refer those who fail to respond and require ventilator support to ICU. If sxs improve, follow up as for moderate.

Life-threatening asthma: silent chest, feeble respiratory effort, cyanosis, hypotension, bradycardia, dysrrythmia, agitation, exhaustion, confusion, LOC, coma. PaO$_2$ < 92%. PFR < 33% predicted. Send immediately to hospital, consult with senior medical staff and refer to ICU. **Treatment** same as for severe acute asthma.

Tx: High flow O$_2.$ Inhaled short-acting B2 agonist via large-volume spacer or O$_2$ driven nebulizer (same as for moderate). Prednisolone 40-50 mg od or IV hydrocortisone 100 mg q 6h (< 1yo 25 mg, 1-5 yo 50 mg, 6-12 yo 100 mg). Monitor response for 15-30 minutes. If poor response, give ipratropium bromide via O$_2$ driven neb 500.

ASTHMA MANAGEMENT IN CHILDREN < 5 YEARS
(BTS/ SIGN Guidelines 2003)

Step 1: short-acting B2 agonist (not more than once a day)

Step 2: regular preventer (short-acting B2 agonist + either regular standard dose inhaled corticosteroid OR if inhaled CS cannot be used, then leukotriene receptor antagonist or theophylline.

Step 3: add on therapy age 2-5: inhaled short acting B2 agonist prn + regular inhaled CS+ leuk receptor antagonist. Age < 2 yo: refer to paeds

Step 4: persistent poor control: refer to paeds/ Stepping down.

Step	Reliever	Addit'l therapies	Further advice/tx
1	inhaled short-acting β2 agonist PRN		Review if high usage.
2	inhaled short-acting β2 agonist PRN	Add inhaled steroid (200μg/day) or add leukotriene receptor antagonist if cannot use steroids.	Usual starting dose 200μg/day divided twice daily.
3	inhaled short-acting β2 agonist PRN	Add inhaled steroid (200-400μg/day)	Consider trial of leukotriene receptor antagonists. If <2 yo, consider proceed to step 4.
4	inhaled short-acting β2 agonist PRN	Add inhaled steroid (400μg/day)	Refer to paediatric specialist.

ASTHMA MANAGEMENT IN SCHOOL CHILDREN & ADULTS
British Thoracic Society and Scottish Intercollegiate Guidelines
Guidelines on Asthma Management. Thorax 2003; 58(Suppl l):i1 i69.

Step	Reliever	Addit'l therapies	Further advice/tx
1 mild intermittent	inhaled short-acting β2 agonist PRN		Review if > 10-12 puffs/day.
2 reg preventer	inhaled short-acting β2 agonist PRN	Add inhaled steroid 200-800µg/day	Titrate to lowest dose. Usual starting dose 400µg/day. Divide dose bd and then od when controlled.
3 add on therapy	as above	as above Add inhaled LABA	Continue LABA, ↑ steroid to 800, if still poorly controlled. If no response to LABA, stop, ↑ steroid to 800µg/day. Consider add 4th drug (leukotriene receptor antagonists, slow release β2 agonist tablets theophylline).
4 persistent poor control	as above	Add inhaled steroid 800µg/day. Add inhaled LABA unless no effect.	Consider trials of ↑ steroids to 2000. Consider trials of 4th drug. Consider specialist referral.
5 continuous or frequent oral steroids	as above	Add inhaled steroid 2000µg/day. Add oral steroids at lowest effective dose.	Consider other txs to minimise oral steroids. Refer to specialist.

Aims: Minimise symptoms during day/ night. No exacerbations. Minimise need for reliever. No physical activity limitation. Normal lung function (FEV1 +/or PEF > 80% predicted or best)

Treatment: Start at level appropriate to asthma severity.

Step up tx as required (prior to starting new tx, recheck compliance, inhaler technique and remove trigger factors; if trials of add-on therapies are ineffective, cease; if trials of increased steroids are ineffective, return to original dose).

Step down treatment levels when control achieved. Review regularly.

Steroid therapy refers to beclomethasone diproprionate via an MDI

Inhaled steroids advised after exacerbations, nocturnal asthma, impaired lung function or with > once daily β2 agonist. 20-50% dose reduction every 3 months.

Systemic side-effects with long-term or frequent oral steroids – monitor BP, review for signs of diabetes and osteoporosis. In children, monitor growth, check for signs of adrenal suppression and screen for cataracts.

The Mx of Atrial Fibrillation (NICE June 2006)

Identification and diagnosis presenting with any of following: SOB, palpitations, syncope/dizziness, chest discomfort, stroke/TIA, **manual pulse palpation** should be performed to assess for an irregular pulse **(AF).**

- **An ECG**, whether symptomatic or not, in whom AF is suspected
  because of an irregular pulse.
- **24h Ambulatory ECG recording** with suspected paroxysmal AF (spontaneously terminates within 7 days, usually within 48 hours) undetected by ECG: with suspected asymptomatic episodes or symptomatic episodes < 24 hours apart.
- **an event recorder ECG** should be used in those with symptomatic episodes > 24 hours apart.

Transthoracic echocardiography (TTE) perform in AF: for whom a baseline echo is important for long-term mx, i.e. younger patients; for whom a rhythm-control strategy that includes cardioversion (electrical or pharmacological) is being considered ; in whom there is a high risk or a suspicion of underlying structural/functional heart disease (i.e. heart failure or murmur) that influences their subsequent mx (i.e. choice of antiarrhythmic drug); in whom refinement of clinical risk stratification for antithrombotic tx is needed.

Transoesophageal echocardiography (TOE) performed in AF: when TTE demonstrates valvular heart disease that warrants further specific assessment; in whom TTE is technically difficult and/or of questionable quality and to r/o ht abnormalities; for whom considering TOE-guided cardioversion.

Cardioversion patients with AF undergoing elective cardioversion. It does not cover those patients with haemodynamic instability following the onset of AF for whom emergency cardioversion may be indicated. See algorithm.

Electrical vs. pharmacological cardioversion In patients with AF without haemodynamic instability for whom cardioversion is indicated. Where AF onset was within 48 h previously, either pharmacological or electrical cardioversion should be performed. For those with more prolonged AF (onset > 48 h previously) electrical cardioversion is the initial treatment.

Pharmacological cardioversion in persistent AF (does not self-terminate, or lasts > 7 d (without cardioversion), where the decision to perform pharmacological cardioversion using an IV antiarrhythmic agent has been made. In the absence of structural heart disease, a Class 1c drug (flecainide or propafenone) is the DOC. In the presence of structural heart disease, amiodarone is the DOC.

Electrical cardioversion with concomitant antiarrhythmic drugs
When AF patients are to undergo elective electrical cardioversion + there is concern about successfully restoring sinus rhythm (i.e. previous failure to cardiovert or early recurrence of AF), concomitant amiodarone or sotalol should be given for at least 4 wks before cardioversion. **TOE-guided elective cardioversion** in patients with AF of > 48 hours' duration if: both TOE-guided cardioversion and conventional cardioversion considered equally effective, where staff and facilities are available, and where a minimal period of precardioversion anticoagulation is indicated due to pt choice or bleeding risks.

<u>**Treatment for Persistent AF - Rate-control vs. rhythm-control**</u> As some patients will satisfy criteria for either an initial rate-control or rhythm-control strategy (i.e., age > 65 but also symptomatic): explain the indications and potential pros and cons of each strategy. Any comorbidities that might indicate one approach rather than the other should be taken into account irrespective of whether a rate-control or a rhythm-control strategy is adopted in patients with persistent AF; use appropriate antithrombotic therapy.
Rate-control strategy preferred initial tx in the following with persistent AF: 65, with CAD, with C/I's to antiarrhythmic drugs, unsuitable for cardioversion (C/I's to anticoagulation, structural heart disease (large l atrium > 5.5 cm, mitral stenosis) that precludes long-term maintenance of sinus rhythm, a long duration of AF (usually > 12 months), a hx of multiple failed attempts at cardioversion and/or relapses, even with concomitant use of, without CHF, antiarrhythmic drugs or non-pharmacological approaches an ongoing but reversible cause of AF (thyrotoxicosis).
Rhythm-control strategy preferred initial tx in the following with persistent AF: those who are symptomatic, younger patients, presenting for 1st time with lone AF, with AF 2° to a treated/corrected precipitant, or with CHF.

Rhythm-control for persistent AF An antiarrhythmic drug is not required to maintain sinus rhythm in patients with persistent AF in whom a precipitant (i.e. chest infection or fever) has been corrected and cardioversion has been performed successfully, providing there are no RFs for recurrence.

In patients with persistent AF who require antiarrhythmic drugs to maintain sinus rhythm and who have structural heart disease: a standard **β-blocker** should be the initial therapy; if ineffective, C/I or not tolerated, use **amiodarone**.

In patients with persistent AF who require antiarrhythmic drugs to maintain sinus rhythm and who do not have structural heart disease (CAD or LV dysfunction): a β-blocker should be the initial tx; if ineffective, C/I or not tolerated give Class Ic agent or sotalol (titrate from 80 mg bd to 240 mg bd); if C/I, use amiodarone.

Antithrombotic therapy for persistent AF Before cardioversion, patients should be maintained on therapeutic anticoagulation with warfarin (INR 2.5, range 2.0 to 3.0) for a min of 3 wks. Following successful cardioversion, patients should remain on warfarin (INR 2.5, range 2.0 to 3.0) for a min of 4 weeks. In persistent AF where cardioversion cannot be postponed for 3 wks: heparin should be given and the cardioversion performed, and warfarin should then be given for a minimum of 4 weeks post cardioversion. Continue anticoagulation for the long term in patients with AF who have undergone cardioversion where there is a high risk of AF recurrence (a history of failed attempts at cardioversion, structural heart disease (MV disease, LV dysfunction or an enlarged LA), a prolonged hx of AF (>12 months), prior recurrences of AF) or where recommended by the stroke risk algorithm. In patients with AF of confirmed duration of < 48 hours undergoing cardioversion, anticoagulation following successful restoration of sinus rhythm is not required. Give atrial flutter patients antithrombotic tx in the same manner as AF.

Treatment for Permanent AF newly dx AF for whom antithrombotic treatment is indicated, treatment should be initiated with minimum delay after the appropriate management of comorbidities.

Rate-control for permanent AF: β-blockers or rate-limiting Ca antagonists are the preferred initial monotx in all patients. Digoxin as monotherapy only in sedentary patients. In patients with perm AF, where monotx is inadequate: to control the HR only during normal activities, give β-blockers or rate-limiting Ca antagonists with digoxin; to control the HR during norm activities and exercise, give rate-limiting Ca antagonists with digoxin.

Antithrombotic treatment for permanent AF: perform and discuss a risk–benefit assessment.

In patients with permanent AF where antithrombotic tx is given to prevent strokes and/or thromboembolism: adjusted-dose warfarin should be given as the most effective tx; adjusted-dose **warfarin should reach a target INR of 2.5 (range 2.0 to 3.0)**; where warfarin is not appropriate, aspirin should be given at 75 to 300 mg/day; where warfarin is appropriate, ASA should not be coadministered with warfarin purely as thromboprophylaxis.

If INR < 5, omit 1 dose, check INR 2-3x a week, resume tx at 10-20% lower dose.

If INR 5-9, omit 1-2 doses, check INR daily, resume tx at 10-20% lower dose when INR reaches patient's target range. If the pt is at high risk of serious bleeding, consider administer vitamin K 2-3mg orally.

If INR > 9, discontinue warfarin temporarily. Consider administer vitamin K 3-5 mg orally. Check INR daily and give additional vitamin K, if INR is not substantially reduced by 24-48h. Resume tx at 20% lower dose when INR reaches pt's target range and monitor INR closely until stable. Consider more frequent routine INR monitoring.

If serious life-threatening bleed (SBP < 90, oliguria, ↓hb, blood transfusion), admit, stop warfarin, give FFP or recombinant Factor VIIa or prothrombin complex concentrate. Administer vitamin K 5-10mg slowly < 1mg/min IV. Monitor INR 6hourly and treatment with repeat vitamin K1 or FFP.

<u>**Treatment for Paroxysmal AF**</u> considers a 'pill-in-the-pocket' (pt self-administers antiarrhythmic drugs only upon the onset of an episode of AF) treatment strategy safe and effective. **Rhythm-control for PAF** Where patients have infrequent paroxysms and few symptoms, or where symptoms are induced by known precipitants (EtOH, caffeine), consider a 'no drug tx' strategy or a 'pill-in-the-pocket' strategy. Initial treatment with standard β-blocker for patients with symptomatic paroxysms (+ /- structural heart disease, including CAD). In patients with PAF and no structural heart disease: where symptomatic suppression is not achieved with β-blockers, give Class Ic agent (flecainide or propafenone) or sotalalol; if not achieved with β-blockers, Class Ic agents or sotalol, use amiodarone or refer for non-pharmacological intervention.

In patients with PAF and CAD: where standard β-blockers do not achieve symptomatic suppression, give sotalol, then try amiodarone or refer. **In patients with PAF with poor LV function**: where standard β-blockers are given as part of the routine mx strategy and adequately suppress paroxysms, no further treatment for paroxysms is needed;

where β-blockers do not adequately suppress paroxysms, consider amiodarone or referral for non-pharma intervention.
Patients on long-term treatment for PAF should be kept under review to assess the need for continued treatment and adverse effects.

Tx strategy for PAF consider a 'pill-in-the-pocket' strategy in those who: have no hx of LV dysfunction, or valvular or ischaemic heart disease; and have a hx of infrequent symptomatic episodes of PAF; and have a SBP > 100 mmHg and a resting HR > 70 bpm; and are able to understand how to, and when to, take meds.
Decision for Antithrombotic tx for PAF should not be based on the frequency or duration of paroxysms (symptomatic or asymptomatic) but on appropriate risk stratification, as for permanent AF.
Tx for Acute Onset AF Acute AF in haemodynamically unstable (life-threatening deterioration) patients- perform emergency electrical cardioversion, irrespective of the duration of the AF. In patients with non-life-threatening haemodynamic instability following the onset of AF: perform elect cardioversion. Use IV amiodarone if delayed in cardioversion. Known WPW syndrome: use flecainide as an alternative for pharmacological cardioversion. AV node-blocking agents (diltiazem, verapamil or digoxin) should not be used. Known permanent AF where haemodynamic instability is caused mainly by a poorly controlled ventricular rate, use a pharmacological rate-control strategy (IV treatment should be with one of the following: β-blockers or rate-limiting calcium antagonists). Amiodarone, where β-blockers or Ca antagonists are C/I or ineffective.
Antithrombotic tx for acute-onset AF who are receiving no, or subtherapeutic anticoagulation tx: in the absence of C/I's, start heparin at initial presentation and continue until a full assessment has been made and anti-thrombotic tx has been started, based on risk stratification. In patients with a confirmed dx of acute AF of recent onset (< 48 hours since onset), use oral anticoagulation if: stable sinus rhythm is not successfully restored within the same 48-hour period following onset of acute AF; or there are factors indicating a high risk of AF recurrence; or it is recommended by the stroke risk algorithm. In patients with acute AF where there is uncertainty over the precise time since onset, use oral anticoagulation, as for persistent AF. In acute AF where pt is haemodynamically unstable, perform any emergency intervention ASAP and the initiation of anticoagulation should not delay emergency intervention.

Antithrombotic treatment with AF who have had a **stroke or TIA** and in asymptomatic patients with AF see algorithm.
Initiating antithrombotic treatment In patients with newly diagnosed AF for whom antithrombotic tx is indicated, initiate such tx with minimal delay after the appropriate mx of comorbidities. Manage uncontrolled hypertensive before start antithrombotic tx in AF who have had an acute stroke. **In patients with AF and an acute stroke**: CT or MRI to r/o cerebral haemorrhage, then start anticoagulant tx after 2 wks; in the presence of bleed, do not give anticoagulant tx; in the presence of a large cerebral infarction, delay anticoagulant therapy.
In patients with AF and an acute TIA: perform **CT or MRI** to r/o recent cerebral infarction or haemorrhage; in the absence of cerebral infarction or haemorrhage, begin anticoagulation ASAP. **Antithrombotic treatment after a stroke or TIA with AF:** Administer warfarin as the most effective thromboprophylactic agent. ASA or dipyridamole should NOT be administered as thromboprophylactic agents unless indicated for the tx of comorbidities or vascular dis. **Tx of post-stroke or post-TIA patients with warfarin** should only begin after tx of relevant comorbidities (hypertension) and assessment of the risk–benefit ratio. **Patients with asymptomatic AF** should receive antithrombotic tx as for symptomatic AF.
Risks of long-term anticoagulation. Assess bleeding risk before tx: > 75 yrs, taking antiplatelets (ASA, clopidogrel) or NSAIDs, on multiple rxs, uncontrolled hypertension, hx of bleeds (peptic ulcer or cerebral), hx of poorly controlled anticoag tx. RFs for stroke and thromboembolism – give thromboprophylaxis. **Monitoring and Referral** Long-term anticoagulation self-monitoring for AF with ed and training for patient or carer; regular review pt's ability to self-manage; reg check equipment for self-monitoring. **Routine FU post successful cardioversion** at 1 and 6 mos to

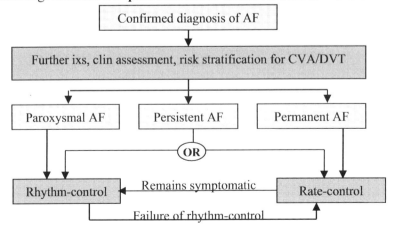

assess maintenance of sinus. At each review, re-assess the need for continued anticoag. At 6 months, if patients remain in sinus and have no other need for hospital FU, discharge from secondary care with an appropriate management plan agreed with GP. Re-evaluate if found at FU to have relapsed into AF. **<u>Consider specialist referral</u>** (pulmonary vein isolation, pacemaker, arrhythmia surgery, AV junction catheter ablation or use of atrial defibrillators) in: failed rx, lone AF, ECG evidence of WPW.

<h2 style="text-align:center">AF care pathway</h2>

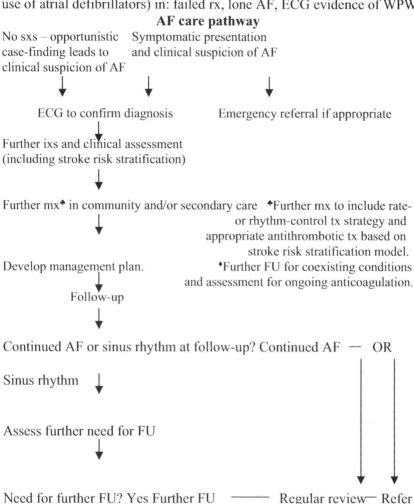

No sxs – opportunistic case-finding leads to clinical suspicion of AF

Symptomatic presentation and clinical suspicion of AF

ECG to confirm diagnosis   Emergency referral if appropriate

Further ixs and clinical assessment (including stroke risk stratification)

Further mx✦ in community and/or secondary care   ✦Further mx to include rate- or rhythm-control tx strategy and appropriate antithrombotic tx based on stroke risk stratification model.

Develop management plan.   ✦Further FU for coexisting conditions and assessment for ongoing anticoagulation.

Follow-up

Continued AF or sinus rhythm at follow-up? Continued AF — OR

Sinus rhythm

Assess further need for FU

Need for further FU? Yes Further FU ——— Regular review— Refer

Haemodynamically unstable AF treatment algorithm

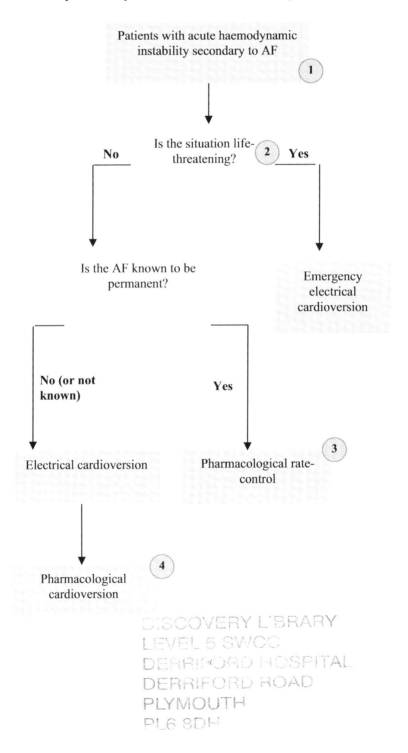

Patients with acute haemodynamic
instability secondary to AF   **1**

No   Is the situation life-
threatening?   **2**   **Yes**

Is the AF known to be
permanent?

Emergency
electrical
cardioversion

**No (or not
known)**

Yes

Electrical cardioversion

Pharmacological rate-
control   **3**

Pharmacological
cardioversion   **4**

Stroke risk stratification algorithm

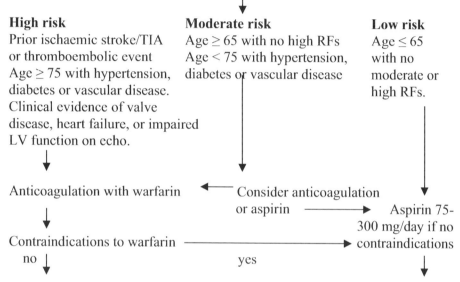

Patients with paroxysmal, persistent or permanent AF

Determine stroke/thromboembolic risk

High risk
Prior ischaemic stroke/TIA
or thromboembolic event
Age ≥ 75 with hypertension,
diabetes or vascular disease.
Clinical evidence of valve
disease, heart failure, or impaired
LV function on echo.

Moderate risk
Age ≥ 65 with no high RFs
Age < 75 with hypertension,
diabetes or vascular disease

Low risk
Age ≤ 65
with no
moderate or
high RFs.

Anticoagulation with warfarin ← Consider anticoagulation
or aspirin → Aspirin 75-
300 mg/day if no

Contraindications to warfarin → contraindications
no | yes

Warfarin target INR 2.5 (range 2-3)   Reassess risk stratification
whenever individual risk factors are reviewed.

ATOPIC ECZEMA (NICE May 2007)

Assess severity, psychological and psychosocial wellbeing and quality of life (everyday activities and sleep) when assessing a child's atopic eczema at each consultation. There is not necessarily a direct relationship between the severity of the atopic eczema and the impact of the atopic eczema on quality of life.

Holistic assessment

	Skin/physical severity		Impact on quality of life and psychosocial wellbeing
Clear	Normal skin, no evidence of active atopic eczema	None	No impact on quality of life
Mild	Areas of dry skin, infrequent itching (with or without small areas of redness)	Mild	Little impact on everyday activities, sleep and psychosocial wellbeing
Moderate	Areas of dry skin, frequent itching, redness (with or without excoriation and localised skin thickening)	Moderate	Moderate impact on everyday activities and psychosocial wellbeing, frequent, disturbed sleep
Severe	Widespread areas of dry skin, incessant itching, redness (with or without excoriation, extensive skin thickening, bleeding, oozing, cracking and alteration of pigmentation)	Severe	Severe limitation of everyday activities and psychosocial functioning, nightly loss of sleep

Identification and mx of trigger factors

When clinically assessing children with atopic eczema, identify potential trigger factors including: irritants, i.e. soaps and detergents (shampoos, bubble baths, shower gels and washing-up liquids); skin infections; contact allergens; food allergens; inhalant allergens. Consider a dx of food allergy in children with atopic eczema who have reacted previously to a food with immediate symptoms, or in infants and young children with moderate or severe atopic eczema that has not been controlled by optimum management, particularly if associated with gut dysmotility (colic, vomiting, altered bowel habit) or FTT.

Treatment - Stepped approach to management

Tailor the treatment step to the severity of the atopic eczema. Emollients should form the basis of atopic eczema mx and should always be used, even when the atopic eczema is clear. Mx can then be stepped up or down, according to the severity of sxs, with the addition of the other txs listed in the table below.

Mild atopic eczema	Moderate atopic eczema	Severe atopic eczema
Emollients	Emollients	Emollients
Mild potency topical corticosteroids	Moderate potency topical corticosteroids	Potent topical corticosteroids
	Topical calcineurin inhibitors	Topical calcineurin inhibitors (tacrolimus and pimecrolimus) 2nd line>2 yo
	Bandages	Bandages (occlusive wet wrap)
		Phototherapy
		Systemic tx (antihistamines) 1/12 trial

Offer info on how to recognise flares of atopic eczema ($\uparrow$ dryness, itching, redness, swelling and general irritability). Give clear instructions on how to manage flares according to the stepped-care plan, and prescribe txs that allow children and their parents or carers to follow this plan.

Emollients

Offer children a choice of unperfumed emollients to use every day for moisturising, washing and bathing. This may include a combo of products or 1 product for all purposes. Leave-on emollients should be prescribed in large quantities (250–500 g weekly) and easily available to use at nursery, pre-school or school.

Topical corticosteroids

The potency should be tailored to the severity of the child's atopic eczema, which may vary according to body site. Use mild potency for the face and neck, except for short-term (3–5 days) use of mod potency for severe flares use moderate or potent preps for short periods only (7–14 days) for flares in vulnerable sites (axillae and groin). Do not use very potent preps in children without specialist dermatological advice.

Treatment for infections

Offer info on how to recognise the sxs and signs of bacterial infection with staphylococcus and/or streptococcus (weeping, pustules, crusts, atopic eczema failing to respond to therapy, rapidly worsening atopic eczema, fever and malaise). Provide clear info on how to access appropriate tx when a child's atopic eczema becomes infected.

Offered info on **how to recognise eczema herpeticum**. Signs of eczema herpeticum are: areas of rapidly worsening, painful eczema; clustered blisters consistent with early-stage cold sores; punched-out erosions (circular, depressed, ulcerated lesions) usually 1–3 mm that are uniform in appearance (these may coalesce to form larger areas of erosion with crusting); possible fever, lethargy or distress. **Recognize picture of eczema herpeticum.**

Education and adherence to therapy

Provide info in verbal and written forms, with practical demonstrations, and should cover: how much of the txs to use; how often to apply txs; when and how to step tx up or down; how to treat infected atopic eczema. This should be reinforced at every consultation, addressing factors that affect adherence.

Indications for referral Referral for specialist dermatologist advice if:

- The dx is, or has become, uncertain.
- Mx has not controlled the atopic eczema satisfactorily based on a subjective assessment by the child, parent or carer (i.e., the child is having 1–2 weeks of flares per month or is reacting adversely to many emollients)
- Atopic eczema on the face has not responded to appropriate treatment.
- The child or parent may benefit from specialist advice on treatment application (i.e., bandaging techniques).
- Suspect contact allergic dermatitis (i.e., persistent atopic eczema or facial, eyelid, hand atopic eczema).
- The atopic eczema is giving rise to significant social or psychological problems for the child or parent/carer (i.e., sleep disturbance, poor school attendance).
- Atopic eczema is associated with severe and recurrent infections (deep abscesses or pneumonia).

Dr Coales's personal anecdote: As a child suffering for years treated ineffectively with paraffin wax emollients, steroids, occlusive bandages, coal tar baths, etc. for severe eczema (swimming exclusion, bleeding, excoriations, crusting, incessant pain and itching, lack of sleep), my GP **finally** referred me to a Consultant Dermatologist at the Royal London Hospital in Whitechapel in 1975. He advised me to stop all tx, use Simple Soap and to ensure I only had direct skin contact with natural fabrics (cotton, wool, silk, linen) and NOT synthetic fabrics made from chemicals (polyester in uniforms and bedding, acrylic which looks like wool, acetate, rayon, viscose, etc.). I told him I was going to grow up and become a doctor much to his disbelief. Within 1 week, years of suffering with severe eczema resolved permanently and now for occasional flare-ups from inadvertently wearing poly mix, I apply 1% hydrocortisone cream and remove all synthetic lining and neck labels. My 3 daughters have all inherited eczema and once spotted as babies, I have followed and given this simple advice to many paeds and adult patients with eczema and their eczema has miraculously resolved for good; no more grocery bags full of emollients to serve as a chemical barrier between polyester and the skin! 30+ years on and I am surprised that NICE does not include this simple advice on management.

Clinical Guidelines for the
Mx of ACUTE LOW BACK PAIN: RCGP 1999

Simple backache (non-specific low back pain) - 90%

20-55 yo; lumbosacral, buttocks, thigh; well pt; 'mechanical' – varies with physical activity and time.
Specialist referral is not required. 90% recover w/n 6/52.
X-rays are not routinely indicated, as soft tissue injuries cannot be detected on x-ray or MRI. Back x-rays have 120x the radiation of a CXR.
Consider psychosocial factors (yellow flags).
Rx regular analgaesia – start with paracetamol, may then substitute NSAIDs and then codydramol. Finally add a short course of a muscle relaxant i.e. baclofen or diazepam. Avoid narcotics.
Do not recommend bed rest. Bed rest for 2-7 days is worse than placebo or ordinary activity.
Advise patients to stay active and resume normal daily activities.
Consider manipulative treatment within the first 6 weeks for patients who are failing to return to normal activities. Refer for reactivation/ rehabilitation if back pain persists > 6/52.

Nerve root pain - < 5%

Specialist referral is not generally required within the first 4/52, provided resolving:

Absent ankle jerk or numbness	Unilateral leg pain worse than LBP
Pain radiates to foot or toes	Pain is worse after coughing

Numbness+paresthaesia are in same direction.
Nerve irritation signs ↓ SLR reproduces leg pain.
Localised neurological signs – limited to one nerve root.
50% recover spontaneously from acute attack within 6/52.
Full recovery expected but recurrence possible.

Red flags for potentially serious spinal pathology (urgent referral < 4/52)

Presentation < 20 yo or > 55 yo	Violent trauma
Constant, progressive, non-mechanical pain	Thoracic pain
PH of CA, drug abuse, HIV or systemic steroids	Structural deformity
Persistent severe restriction of lumbar flexion	Widespread neurology

Systemically unwell, weight loss
Note: MRI has a high FP rate and is abnormal in everyone > 30 yo, will show disk protrusion. By age 50, we all have lumbar spondylosis.

Cauda equina syndrome (immediately refer - gait disturbance, saddle anaesthesia, sphincter disturbance).

Yellow Flags (psychosocial risk factors): Kendall et al, 1997.

A belief that back pain is harmful or potentially severely disabling.
Fear-avoidance behaviour and reduced activity levels. Have you had time off work in the past with back pain? When do you think you will return to work?
Tendency to low mood and withdrawal from social interaction. Expectation of passive treatment (s) rather than a belief that active participation will help.
Patients are at risk of chronic low back pain and poor prognosis.
Consider specialised psychological referrals for those with psychopathology.

Vertebral fractures - < 5% - Assess RFs: early menopause, previous fracture, prolonged steroids
Inflammatory conditions – 1% i.e. ankylosing spondylitis (early 20s M, AM stiffness, pain or after sitting for 1hour in front of the television)
CA/ infection - < 1% -painless Paget's dis. Metastatic pain is continuous, unremitting, throbbing nad nocturnal.

BENEFITS (WELFARE)

1.   Free Prescriptions   < 16 yo; 16-18 yo in F/T education; 60 or >; pregnant or had a baby

in the last yr (FP92); war or MoD pensioner medical exemptions (Addison's disease, DM, epilepsy, fistulae, hypopituitarism, myasthaenia gravis, myxoedema)

entitled to a prepayment certificate (FP96)

NHS low income scheme (HC2 cert)

receiving benefits from income support, family credit, DWA (disability work allowance), job-seeker's allowance

2.   Disability and Handicap

Attendance Allowance   > 65 yo, severe mental or physical disability; lasting > 6/12, day care or day + night rate, tax free, non-means tested; supervision to prevent injury and help required to perform bodily functions

Disability living allowance   > 5-yo and < 65-yo; requires care + immobility, (unable to walk) for 3/12, lasting > 6/12.
DWP doctor must assess blindness, etc.

Disability working allowance low income, > 16 yo, 16h/ wk, receiving one of the disability allowances; tax-free BUT income-related and means-tested

Incapacity Benefit   paid at 3 rates (short-term up to 28 wks at lower rate, higher rate from 28 wks –1 year, long-term rate after 1 year); if one is unfit to work on medical grounds for > 28 wks (issue med 4 after 28 wks) and has paid NI contributions; 'All- work test' – objective assessment of mental and physical capability for work

Invalid Care Allowance   carers with dependants, handicapped; 16-64 yrs; 35h/wk as carer, <£50/wk earnings; depends on AA, DLA, constant attendant allowance

Severe Disability Allowance 80% disabled or unable to work < 20th
birthday; no NI contributions; unable to work
for > 28 wks, 16-65 yo

3.   Low Income and Other Benefits

Child benefit < 16 yo, tax free; not means tested

Disabled Persons Tax Credit work > 16h/wk; means-tested

Incapacity benefit not entitled to statutory sick pay; self-employed

Income support =>18yo, < 16 h/wk; means tested; savings £3k

Maternity incapacity benefit in last 6/52 and 2/52 postpartum

Six Social fund payments budget loan; crisis loan; community care grant;
cold weather payments, funeral expenses;
maternity grant

Statutory maternity allowance worked at least 26 wks of 66 wks prior
to due date and paid NI for 8/52 prior. 18 wks
pay; apply for MATB1 > 20/40

Statutory sick pay 16- 65 yo; unable to work between 4/7 and 28
weeks

Visually handicapped ≥3/60 blind, < 6/60 partially sighted, cheaper tv
licence, parking + travel fare concessions

Deafness SS arrange doorbells, flashing alarms, travel
concessions

Working Family's Tax Credit work > 16h/wk with children; means tested

Bereavement Allowance 52 weeks

EUROPEAN HEALTH INSURANCE CARD
www.ehic.org.uk

- any medical treatment that becomes necessary during your stay due to illness or accident.
- access to reduced-cost or free medical treatment from state healthcare providers.
- allows you to be treated on the same basis as a resident of the country you are visiting i.e. you may have to pay a patient contribution (a co-payment). You may be able to seek reimbursement for this when you are back in the UK if unable to do so in the other country.
- treatment of a chronic or pre-existing medical condition that becomes necessary during your visit.
- routine maternity care, (not solely illness or accident) provided the reason for your visit is not specifically to give birth. Covers the cost of all

medical treatment, for mother and baby, linked to the birth where that occurs unexpectedly.
• includes the provision of oxygen, renal dialysis and routine medical care.

ADMIN/ PATIENT LIST REMOVAL (new GMS contract – para 192-201)

GPs have the right to request any patient be removed. Where a practice has reasonable grounds to remove the patient, it must inform the PCT in writing, and must notify the patient in writing of the reasons for removal. However, where the practice believes that it is not appropriate to give specific reasons it is sufficient to state that there has been a breakdown in the relationship between the practice and the patient. Care should be taken to ensure that the reasons given are factual and that the tone of the letter is polite and suitably informative. The removal will not take effect until the 8th day after the request is received by the authority or 8 days after treatment at intervals of < 7 days ceases unless the patient is accepted by, allocated or assigned to another GP sooner than this. The patient is always notified by the PCT. Patient complaints must be acknowledged within 2 days and responded to within 10 days (PCT 20 days).

BNF SYMBOLS

Black Δ symbol newly licensed meds monitored by MHRA. Report all suspected reactions through Yellow card scheme. **Yellow card scheme**: doctors, dentists, coroners, nurses, pharmacists and self-reporting patients may report adverse reactions to www.yellowcard.gov.uk or by phone/post to Yellow Card Centres or MHRA.
PGD (Patient group direction) is a written direction relating to the supply and administration of a licensed rx-only medication signed by a doctor, dentist or pharmacist.
PoM (Prescription only meds) available only on rx issued by an appropriate practitioner. Know examples.
Prescription writing: Avoid unnecessary use of decimal points, i.e. 3 mg NOT 3.0 mg. May write 0.5 mL but no cc. Write quantity in grams as 1g. Quantities < 1g write in mgs, i.e. 500 mgs NOT 0.5g. Quantities < 1mg write as 100 micrograms and NOT 0.1mg. Micrograms and nanograms should NOT be abbreviated.

Breast Screening Programme (www.cancerscreening.nhs.uk)

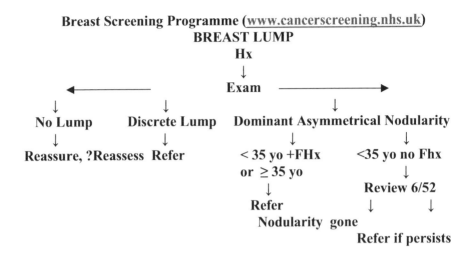

Which BURNS need referral? www.cks.library.nhs.uk

Immediately refer people with burns fulfilling the following criteria:
All deep dermal and full-thickness burns. Superficial dermal burns of more than 10% total burn surface area (TBSA) in adults (≥ 16 years old), or more than 5% TBSA in children (< 16 years old). Superficial dermal burns involving the face, hands, feet, perineum, genitalia, or any flexure (particularly the neck or axilla), or circumferential burns of the limbs, torso, or neck. Any significant infection, septic episode, or suggestion of toxic-shock-like illness. Any inhalation injury. Any electrical or chemical burn injury. Suspicion of NAI or deliberate self-harm. Burns associated with other injuries (e.g. crush injuries, fractures, head injury, penetrating injury).

Consider referral for people with burns fulfilling the following criteria (clinical judgement should be used, taking into account factors including the size of the burn, the patient's level of discomfort and social circumstances, and the experience and confidence of the primary care team in managing burns): Superficial dermal burns other than those mentioned above.
Children under 5 years or adults over 60 years of age. Patients with coexisting medical problems (e.g. cardiac, respiratory, or hepatic disease, or diabetes) or pt who are immunosuppressed or who are pregnant. Patients with burns who may require admission for social reasons, pain control, or if dressings are difficult to manage. Consider non-acute referral to a local plastic surgery unit for any wound that has not healed 14 days after injury. If there is uncertainty whether referral is appropriate, seek specialist advice.

BURNS:

Superficial or 1st-degree burn - the epidermis is damaged. The skin becomes red, slightly swollen and painful. Heals by itself in less than one week without scar. No barrier. Sunburn.

2nd degree: Superficial 2nd degree: heat destroys entire epidermis and < 1/3 of upper dermis (partial thickness). Healing 1-2 weeks. No scar. Initial severe pain. Low risk for infection. Debride. Large blisters can be debrided off if using a temporary skin substitute or left intact for a few days. Topical antibiotics (cream based silver sulfadiazine) impede healing and are only used if infection risk is high. If hot water scald, the painful wound looks pink and wet as blisters disrupt after 12-24 hours. Antibiotic creams slow healing rate. Clean, remove small blisters; apply grease gauze and soft gauze dressing (occlusion, absorbent dressing, changed daily). On the face, perineum, apply bacitracin or neomycin ointment, several times daily. Use a water-soluble topical antibiotic if the wound is grossly contaminated or if unsure if the wound is superficial or deep. No need for prophylactic systemic antibiotics.

Mid-2nd degree (mid-partial thickness burn): ½ dermis destroyed (hot liquids 5-10 seconds or flash flame with no direct contact). 2-4 weeks healing. Less blood supply. Less intense pain (nerve partially destroyed) as superficial. Red blisters, less wet. If heals in 2 weeks, then minimal to no scarring. > 3 wks, scarring will occur, ↑dark patients. **KELOIDS most common in upper arm deltoid, back and sternum in black patients.** In 6 to 60 years, without diabetes, chronic illness, treatment is grease gauze, an occlusive dressing and a topical antibiotic ointment. In the very young, and very old, or those with chronic illness, contaminated wounds or perineal wounds, use silver sulfadiazine or silver dressing with closed dressing technique. Temporary skin substitute can ↑ healing, protect the wound and ↓ pain.

Mid-dermal burn: **transfer to burn centre if location is feet or size (> 15% TBS).** Too big to use cold dressings except for a very brief initial period. Debride loose tissue. Grease gauze, topical antibiotic ointment or silver dressing with closed gauze dressing, consider temporary skin substitute for wound closure. If hand, then cold water to control pain, cleanse, closed technique grease gauze plus dressing. Antibiotic ointment optional, silver cream not needed. Apply dressing to allow for hand mobility.

Deep 2nd degree (deep partial thickness) burn (flames): Most of skin is destroyed except for small amount of remaining dermis. Dry, white or charred skin. Compromised blood flow and a layer of dead dermis or eschar adheres to the wound surface. Minimal pain as the nerves are destroyed. Distinguish a deep dermal from a full thickness (3°) by visualization. The presence of sensation to touch indicates the burn is a deep partial injury. 4-10 weeks to heal. Since the new epidermis is very thin and not adhered well to dermis, wound breakdown is common. Excision and grafting is the preferred. Dense scarring is usually seen if the wound is allowed to heal primarily. Involves majority of the inner dermal layer. High risk of infection. Severe scar. Readily converts to a full thickness burn.

Admit if > 2% due to need for early grafting. Transfer to Burn Center based on Transfer Criteria. Gentle washing with antibacterial soap. Silver sulfadiazine using a closed dressing or silver impregnated dressing. Cold is not beneficial once the burning stops as pain is minimal.

Mid to Deep Hand Burn treated with silver cream. Silver is constantly released over a 3-5 day period resulting in excellent infection control but with fewer dressing changes

Full thickness or 3rd-degree burn - the epidermis, dermis and subcutis are damaged. The affected skin will have burned away and the tissue appears pale or blackened. **3rd degree (full thickness) burn (direct exposure with a flame, contact with hot grease, tar or caustic chemicals):** Both layers of skin are completely destroyed; no cells to heal. Any significant burn will require skin grafting. Small burns will heal with scar. Initially the avascular burn tissue is waxy white. If the burn produces char or extends into the fat as with prolonged contact with a flame source, appears leathery brown or black with surface coagulation veins. Painless and has a coarse non-pliable texture.

Except for a very small wound, e.g. 2x2 inches, the burn wound will require excision and a skin graft. Full thickness (3° degree) burn of the arm and chest. Presence of char. Painless due to loss of nerve endings. Transfer to Burn Center based on Transfer Criteria (size, depth and patient's age). No need to perform burn care if transfer is immediate). Gentle washing with antibacterial soap. Silver sulfadiazine using a closed dressing bd or use of a silver impregnated dressing. Cold is not beneficial once the burning has stopped as pain is minimal.

Hot water immersion scalds: Prolonged hot H_2O contact produces a deep burn in the elderly or in young children who cannot escape hot H_2O. Consider forced immersion or abuse. The water vehicle transmits heat to tissues 20 x > than air; the tissue is injured deeper than with a flash flame of short exposure. However, the water T° is usually not hot enough to immediately coagulate vessels, so the wound looks red like a superficial burn but soft tissue injury including nerves can be severe and the burn can be very deep. Burn size and area are indicators that pt needs care in a Burn Center. Notify Social Services!

WWW.CANCERRESEARCHUK.ORG

Figure 4.1: The ten most common cancers, females, UK, 2006

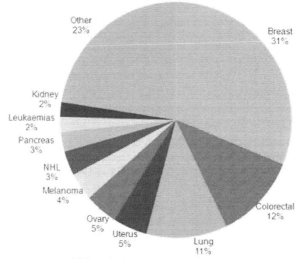

All female cancers (exc NMSC): 146,378

Figure 3.1: The 10 most common cancers in males, UK, 2006

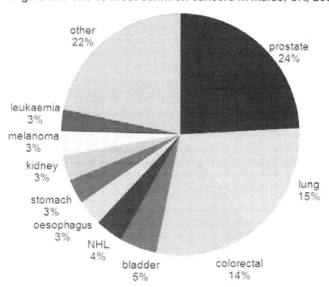

All male cancers (exc NMSC): 147,223

Cancer Referral Guidelines for Suspected Cancer (NICE June 2005) - Urgent Referrals
Breast Cancer:

- Discrete, hard lump with fixation, +/- skin tethering.
- Age < 30: benign lumps (fibroadenoma) or breast pain + no palpable abnormality = non-urgent.
- Age < 30: lump that enlarges, or is fixed and hard, or reason for concern i.e. FH.
- Age ≥ 30 with a discrete mass persisting after next period or presenting after menopause
- Any one of: spontaneous unilateral bloody nipple, unilateral eczematous skin or nipple change not responding to tx; recent nipple distortion; previous histologically confirmed breast CA + lump or suspicious sxs.
- Men ≥ 50 with unilateral, firm subareolar mass +/- nipple distortion or associated skin changes.

Lung Cancer is the # 2 cancer for men and the # 3 cancer for women in the UK after breast and bowel:
- Persistent haemoptysis in smokers or ex-smokers ≥ 40 yrs; SVC obstruction; Stridor.
  Any of the following:
- Haemoptysis
- Unexplained persistent (> 3wks): chest/ shoulder pain, dyspnoea, wt loss, chest signs, hoarseness, finger clubbing, cervical/ SC LNs, cough +/- any of the above, features suggestive of mets from lung CA = CXR (report back in 5/7). Urgent referral: Normal CXR but high suspicion of CA OR CXR suggestive of lung CA (pleural consolidation, pleural effusion).
- Patients presenting with the above sxs or signs + the following RFs (current or ex-smoker, COPD, exposure to asbestos, h/o CA (head+neck) may be referred earlier for CXR OR to a specialist.

Skin Cancer:
- Melanoma: Pigmented lesion with:
- Major features: change in size, irregular shape, irregular colour
- Minor features: largest diameter 7 mm or >, inflammation, oozing, change in sensation.

- Major features (2 points), minor features (1 point). Suspicion is greater for 3 patients or more.
- Squamous cell CA: Nonhealing lesions > 1 cm with significant induration on palpation, common on face, scalp or back of hand with a documented expansion over 8 wks.
- Histological dx of SqCCa
- PHx of transplant and a new or growing skin lesion
- Persistent of slowly evolving unresponsive skin conditions in which the dx is uncertain and CA is a possibility = referral to dermatology.
- Suspicion of basal cell CA (usually face) = non-urgent referral

Gynaecological cancers (urgent referral):
- persistent IMB + negative pelvic exam (consider urgent referral)
- Clinical suspicious features of the cervix
- PMB and not on HRT OR persistent unexplained PMB after having stopped HRT for 6 wks OR PMB while taking tamoxifen
- Vulval sxs should prompt a vulval exam. Unexplained vulval lump.
- Vulval bleeding due to ulceration.
- Menstrual changes, IMB, postcoital, PMB or PV discharge prompts full pelvic exam + cx speculum. If abdo or pelvic mass (not uterine fibroids, GI or urological origin), arrange urgent U/S +/or refer urgently.
- Any unexplained abdo sxs, abdo palpation should be undertaken + pelvic exam considered.
- Vulval pruritis or pain (treat and wait) with active FU. If persists, urgent or non-urgent referral depends on degree of concern about CA.

Upper GI cancer:
- Dysphagia
- Unexplained worsening dyspepsia + any 1 of: known dysplasia, atrophic gastritis or intestinal metaplasia, Barrett's oesophagus, peptic ulcer surgery > 20 yrs ago.
- Any one of: unexplained upper abdo pain + wt loss, upper abdo mass without dyspepsia, obstructive jaundice.
- Without dyspepsia, with any one of: unexplained weight loss, iron deficiency anaemia, persistent vomiting + wt loss.
- Dyspepsia with any of the following risk factors = urgent referral or endoscopy.

- chronic GI bleeding   epigastric mass
- persistent vomiting   suspicious barium meal
- progressive unintentional weight loss   iron deficiency anaemia
- in a pt ≥ 55 years with unexplained and persistent recent-onset dyspepsia alone = urgent endoscopy

Lower GI Cancer (Bowel is the #2 cancer for women and #3 cancer for men in the UK):

- equivocal sxs but patient is not unduly anxious: treat, watch and wait
- unexplained sxs related to LGI tract: perform DRE, if palpable rectal mass (intraluminal and not pelvic) = refer urgently
- 40+ yo rectal bleeding + change in bowel habit (looser stools, +/or ↑ stool frequency) x > 6/52.
- 60+ yo rectal bleeding x 6/52 without change in bowel habit or anal sxs OR change in bowel habit x 6 wks without rectal bleeding.
- RLQ abdo mass c/w involvement of the large bowel.
- Unexplained Fe deficiency anaemia + hb ≤ 11g/100ml in M and hb ≤ 10g/100ml in non-menstruating F

Haematological Cancers:

- Combination of: fatigue, drenching night sweats, F, wt loss, generalised pruritis, breathlessness, bruising, bleeding, recurrent infections, bone pain, EtOH-induced pain, abdo pain, LN, SPM = full exam, including FBC+blood film = If reported as acute leukaemia, refer immediately.
- Unexplained LN or persistent unexplained fatigue = Ix with blood film, FBC, ESR/ PV/ CRP.
- Unexplained bleed, bruising, purpura or sxs of anaemia = Ix: blood film, FBC, ESR/ PV/CRP, clotting.
- Persistent unexplained bone pain = Ix (FBC, urea, lytes, LFT, bone profile, x-ray, PSA, ESR/PV/ CRP).
- Following features of lymphadenopathy: persistence for ≥ 6 wks, LNs ↑ in size, LN > 2cm, widespread nature, associated splenomegaly, night sweats or weight loss = Ix, FU +/or refer.
- Persistent unexplained splenomegaly = refer urgently.
- Spinal cord compression or renal failure suspected of being cause by myeloma = refer immediately.

Urological Cancer (prostate) is the # 1 cancer for men in the UK:
- If hard irregular prostate on DRE
- High PSA levels
- Normal prostate on DRE, +/- lower urinary tract sxs but PSA↑.
- Asymptomatic with borderline PSA. Repeat PSA after intervals of 1-3 months. Repeat shows level ↑.
  Unexplained: Erectile dysfunction, haematuria, LBP, bone pain, weight loss or prostate CA sxs = DRE+PSA (first rule out UTI)

Urological Cancer (renal):
- Painless macroscopic haematuria
- Macroscopic haematuria with sxs suggestive of UTI, investigate and treat accordingly before consider referral. If UTI not confirmed = urgent referral
- Age ≥ 40 with persistent or recurrent UTI with haematuria
- Age ≥ 50 with unexplained microscopic haematuria
- Abdo mass identified clinically or on imaging thought to arise from urinary tract
- Swelling or mass in the body of the testis.
- Age < 50 with microscopic haematuria should have proteinuria testing + creatinine. If proteinuria or ↑ creatinine = refer to renal physician. If no proteinuria and creatinine normal = non-urgent referral to urologist.

Head and Neck Cancers (urgent referral to H+N team):
- Hoarseness ≥ 3 wks with negative CXR
- Any of the following:
- Unexplained lump in neck recently appeared or changed over 3-6 wks
- Unexplained ulceration of oral mucosa or mass > 3 wks
- Unexplained persistent swelling in the parotid or submandibular gland
- Unexplained persistent sore or painful throat
- Unexplained pain in H+N > 4 wks assoc with otalgia but with normal otoscopy
- Unexplained red and white patches of the oral mucosa + pain, swelling or bleeding
- Hoarseness ≥ 3 wks = urgent CXR. If CXR suggestive of CA, refer urgently to respiratory specialist
- Unexplained tooth mobility > 3 wks = urgent referral to dentist

- Unexplained red and white patches of oral mucosa without pain, swelling or bleed = refer non-urgently
- Oral lichen planus - monitor as part of routine dental care.

Head and Neck Cancers - thyroid (urgent referral):

- A thyroid swelling assoc with any of the following: a solitary nodule increasing in size; h/o neck irradiation; FH of endocrine CA; unexplained hoarseness or voice changes; cervical LNs; pre-pubertal pt; ≥ 65 yo
- Persistent sxs or signs related to the oral cavity in whom a definitive dx of a benign lesion cannot be made should be referred or FU until disappears. If persists > 6 wks, refer urgently.
  Immediate referral: Sxs of tracheal compression including stridor due to thyroid swelling
  Non-urgent referral:
- Thyroid swelling without stridor or any listed features should have tfts. Swelling + normal TFTs.
- Thyroid swelling without stridor or any listed feat. Swelling + abnormal TFTs. Non-urgent to endocrinologist.

Brain and CNS Cancers (urgent referral):

- CNS sxs including: progressive neurological deficit, new onset seizures, HA's, mental changes, CN palsy, unilateral SNHL in whom brain tumour is suspected. Suspected recent onset seizures (detailed hx, PE + developmental assessment should be carried out)
- Rapid progression of any of the following: subacute focal neurological deficit; unexplained cognitive impairment or behavioural disturbance or slowness or a combination; personality changes confirmed by witnesses with no reasonable explanation.
- Recent onset HA's with sxs of ↑ ICP (vomit, drowsy, postural-related HA's with pulse-synchronous tinnitus) or focal/ non-focal neurological sxs (blackout, change in personality or memory)
- New, qualitatively different unexplained HA becoming progressively severe
  New unexplained HA or neurological sxs prompts neurological exam (absence of papilloedema does not r/o possibility of brain tumour)
  Recent onset unexplained HA ≥ 1 mo without ICP features = discuss with specialist or non-urgent referral.

Bone and soft-tissue cancers (urgent referral):

- ↑ unexplained or persistent bone pain or tenderness, particularly pain at rest (and especially if not in the joint) or an unexplained limp = urgently Ix. In older people mets, myeloma or lymphoma and sarcoma should be considered. If possible = refer urgently.
- Suspected spontaneous fracture = immediate x-ray = if suggest CA, refer urgently. If normal but persisting sxs, FU +/or repeat x-ray or bone function tests or request referral
- Palpable lump that is either > 5cm, deep to fascia, fixed or immobile, painful, ↑ in size, a recurrence after previous excision
- HIV, consider Kaposi's sarcoma and refer if suspect.

Children's cancers:

Brain tumour:

- ↓ LOC = immediate admission
- persistent HA requires full neuro exam or urgent referral
- Age > 2 yrs and any 1 of neuro sxs and signs: new onset seizures, CN abnormalities, visual disturb, gait abnormal, motor or sensory signs, unexplained deteriorating school performance or developmental milestones, unexplained behavioural and/ or mood disturbances = refer urgently or immediately.

Immediate referral

- > 2 yo with HA and vomiting causing early waking or occur on waking
- < 2 yo + any 1 of: new onset seizures, bulging fontanelles, exterior attacks, persistent vomiting
- < 2 yo. A CNS tumour is suggested by abnormal ↑ in head size, arrest or regression of motor development, altered behaviour, lack of visual following, poor feeding. FTT, squint (urgency contingent on other features) = refer urgently

Leukaemia and lymphoma

- LNs with ≥ 1 of (in absence of local infection): axillary or SC nodes; nontender, firm/hard; >2cm, LNs assoc with signed of generalised ill health, fever +/or weight loss
- ≥ 1 of (alone or with SOB): pallor, fatigue, unexplained irritability or F, persistent or recurrent UTIs, general LN, persistent or unexplained bone

pain, unexplained bruising = Ix with blood film/ FBC. If acute leukemia, refer urgently.

- Immediate referral - mediastinal or hilar mass on CXR; unexplained petechiae or hepatosplenomegaly

Neuroblastoma and Wilm's tumours (urgent referral)

- Any 1 of: pallor; persistent or unexplained bone pain; fatigue; unexplained irritability or F; persistent or recurrent URTI; general LNs; unexplained bruise = FBC = if results indicate anaemia, consider neuroblastoma
- Symptoms that could be explained by neuroblastoma: abdo exam and/ or urgent U/S should be undertaken and FBC/ CXR should be considered. Any mass identified = urgent referral.
- Any 1 of: unexplained back pain, leg weakness, proptosis, unexplained urinary retention.
- Progressive abdo distension = abdo exam. Mass found = immediate referral.
- Haematuria
- Infants < 1 yr with abdo or thoracic mass should be referred immediately.

Bone tumours, sarcoma and retinoblastoma (urgent referral)

- Unexplained mass: deep to fascia; non-tender; progressive enlargement; assoc regional LN enlargement; or size > 2cm.
- Proptosis, persistent unexplained unilateral nasal obstruction +/- discharge/ bleeding, aural polyps/ discharge, urinary retention, scrotal swelling, blood stained vaginal discharge = consider sarcoma.
- Visual problems and fhx of retinoblastoma alert diagnosis.
- Persistent localised bone pain and/ or swelling = x-ray. If suggests osteosarcoma, refer urgently.
- Rest pain, back pain or unexplained limp. Discuss with paeds +/ or x-ray. Paeds advises refer urgently.
- White pupillary reflex, new squint, change in VA with suspicion of CA.

CERVICAL SMEAR SCREENING

- According to the Cancer Research UK NHS Cervical Screening Programme, F < 25 should not be screened and recommends 3-yrly screening between 25 and 49 and 5-yrly screening between 50 and 64.
- In the US, at 2-yrly interval, as studies show cervical CAs are missed if screened at later intervals.

Borderline atypia or dyskaryosis (Borderline CIN)

- repeat smear at 6 months; refer for colposcopy if there are 2 or 3 borderline smears at 6-monthly intervals

Mild dyskaryosis (Grade 1 CIN)

- check local policy; contentious whether to repeat smear at 6 mos or refer immediately for colposcopy as there is a 30% risk of the pt already having CIN 2 or 3 changes.

Mod and severe dyskaryosis (Grades 2 and 3 CIN)
refer immediately for colposcopy

- Study in NZ: 1/3 of untreated F with CIN 3 developed cervical CA over a 20-yr period.
- Glycogen in healthy cells takes up Lugol's iodine so normal cells are iodine-stained. CA cells are devoid of glycogen so do not take up iodine.

CHEST EXAM (LUNG SOUNDS)

Apical fine creps - ankylosing spondylitis, pulmonary fibrosis, tuberculosis
Basal fine creps - fibrosing alveolitis, LVF
Localised fine creps – pneumonia
Basal coarse creps - bronchiectasis, hypostasis
Scattered coarse creps - bronchiectasis, chronic bronchitis
Early and mid inspiratory creps - proximal airways; **Late inspiratory** - alveoli; **Expiratory creps** - airflow obstruction
Dullness, ↑ clear vocal resonance/ fremitis '99', bronchial breath sounds heard far from larger airways – consolidation
Pleural rub – sand paper, pleurisy, pulmonary infarction, pneumonia, pleural CA or spontaneous pneumothorax
Resonance – hyper expanded chest or pneumothorax
Rhonchi – airway secretions
Stony dullness, ↓ vocal fremitis – pleural effusion
↓ vesicular breath sounds – COAD, pneumonia and absent (pneumothorax, mucus plugging)
Wheezes – bronchospasm, airway oedema, CHF, endobronchial CA, secretions

CHILD DEVELOPMENT MILESTONES

Seminars in child and adolescent psychiatry 2nd ed Simon Gowers.
Royal College of Psychiatrists '05

Age	Motor	Speech	Vision and Hearing	Social development
6-8 wks	Moro, grasp, vocalises, follow objects 1m away; startles, smiles at mum (4-6/52)			
12 wks/3mos	prone, no grasp reflex, head held up for long periods, talks a lot, turn head to sounds on the same level with the ear, follow dangling toy, squeals w joy, discriminates smile			
5 months	holds head steady, objects taken to mouth, enjoys vocal play, smiles at mirror image			
6 months	sits with support, rolls over prone to supine, transfers cube hand to hand, crude touch, palmar grasp of cube, grasp rattle, double syllable sounds (mumum, dada), localises sound 45 cm lateral to either ear, may show 'stranger shyness', person preference.			
7 months	feeds self a biscuit, turns head to sound below the level of the ear			
9-10 months	sits without support, crawls on abdomen, babbles, afraid of strangers, fine pincer grasp			
12 months	stands holding furniture, stands alone for 1-2 sec, 2-3 word babble, walks with hand held			
18 months	walk solo, stairs with rail, 10-12 words, asks for potty, uses spoon, 3-4 cube tower, jump with 2 feet, toss ball, uses cup with both hands, feeds self with spoon, 5-50 words 1-2 yo			
2 years	6-7 cube tower, kicks a ball, plays with children alongside, 2-3 word sentences, walks up/down 2 feet per step, runs, dry by day, parallel play, uses 50-300 different words			
3 years	knows age and sex, joins 3 words into a sentence, plays with other children, 1 foot upstairs and 2 feet per step down, knows 2 colours, copies circle, imitate cross and draw man on request, 9 cube tower, asks Qs, undresses solo, imaginary friend, jump on spot			
4 years	knows first and last name, can copy a cross, catch a ball, rides a tricycle, dresses with help, 1foot stairs, skips on 1 foot, imitate gate with cubes, attends to own toilet needs			
5 years	count 10-12 objects, knows 3-4 colours, hops on one foot, skips on both feet, has own set of friends, draws a man and copies a			

a triangle, fluent speech, number of fingers, knows R from L, give age, copies a diamond, dresses and undresses alone.

CHILDHOOD IMMS (NHS Imms, Jan 2008)

2 months	DTaP (diphtheria, tetanus, pertussis)/ IPV (polio)/ Hib + Pneumococcal conjugate vaccine (PCV)
3 months	DTaP/ IPV/ Hib + Men C
4 months	DTaP/ IPV/ Hib + Men C + PCV
12 months	Hib/ Men C
13 months	MMR (thigh 23 G blue needle)/ PCV
3y4mo or after	DTaP/IPV or dTaP/IPV (arm) + MMR
13-18yo	Td/IPV

CHILDHOOD ILLNESSES

Chickenpox incubation 14-16 days; short or no interval between onset and rash, rash lesion of different stages – spots in crops → macule → papule → vesicle. Complications: eczema herpeticum, encephalitis, pneumonia

Erythema infectiosium (Parvovirus, 5th disease) mild fever, slapped cheek (erythematous maculopapular rash), reticular lacy rash on trunk and limbs, duration 4-7 days

Hand, foot and mouth (Coxsackie) mild F, oral blisters, red vesicles on palms + soles, duration 4-7 days

Kawasaki's Disease 9-12 mo old; < 5 y; T > 5d, skin peels at 2-3 wks, cracked lips, swollen LNs, conjunctivitis; echo r/o coronary artery aneurysm, rx ASA + gammaglobulin <10 days

Measles incubation 10-14 days; early – coryzal symptoms, conjunctivitis, fever, later - buccal Koplik spots , 4 days between onset and florid maculopapular rash on face, behind ear, chest. Complications: bronchopneumonia, encephalitis, subacute sclerosing panencephalitis. Duration 10 ds. Antibiotics for 2° infections.

Mumps incubation 16-21 days; fever, malaise, parotid or submandibular gland enlargement. Complications: aseptic meningitis, epidydimo-orchitis, pancreatitits.

Overdoses Common cause for < 6 year olds is iron OD. Moderate 40mg/kg. Lethal 60 mg/kg. Rx: desferoxamine, gastric lavage, fluids.

Roseola infantum < 2yo, fever, sore throat, after 3-4 days faint erythematous macular rash when fever goes does, much better, duration 4-7 days

Rubella/ German incubation 14-21 days; mild, suboccipital LNs, short or no interval

Rubella/ German Measles between disease onset and pink macular rash; transient rash, 15 yo
Complications: birth defects in pregnancy, arthritis in teens. Duration 10 ds.
Scarlet Fever Incubation 2-4 days; 1-2 day interval; long infectivity period shorten by tx, scarlet facial flushing, coated strawberry tongue, truncal rash, non-exudative tonsillitis, circumoral pallor. Complications: acute GN, rheum fever. Rx pen 10/7.
Typhoid Incubation 3/7-3/52. 7-14 days between onset and rash; malaise, fever, cough, nose bleeds, bruising, truncal rose spots, splenomegaly. Coma. Adverse reaction to vaccine increases in > 35 yo, given 3yrly

Fever in CHILDREN < 5 yo (NICE May 2007)

Detection of fever (do not use oral, rectal and forehead thermometers)
Ages 4 wks to 5 yrs, measure body temp by electronic thermometer or chemical dot thermometer in the axilla, or infra-red tympanic thermometer. Reported parental perception of a fever should be taken seriously.
Clinical assessment of the child with fever
Children with feverish illness should be assessed using the traffic light system.
Measure and record T, HR, RR + capillary refill time as part of routine assessment of child with fever.
Children with the following sxs or signs should be recognised as being in a
high-risk for serious illness:
- unable to rouse or if roused does not stay awake moderate/ severe chest indrawing grunting
- weak, high-pitched or continuous cry RR> 60 breaths/ min
- pale/mottled/blue/ashen bulging fontanelle
- reduced skin turgor appearing ill to a
- bile-stained vomiting healthcare professional

Children with any of the following sxs should be recognised as being in at least
an intermediate-risk:
- wakes only with prolonged stimulation ↓ urine output
- ↓ activity a new lump > 2 cm
- poor feeding in infants pallor reported by
- dry mucous membranes parent or carer
- not responding normally to social cues/no smile nasal flaring

Children who have all of the following and none of the high or intermed features, are low-risk:
- strong cry or not crying normal colour of skin, lips and tongue

- content/smiles
- stays awake
- normal response to social cues.

normal skin and eyes
moist mucous membranes

Be aware that a ↑ HR can be a sign of serious illness, particularly septic shock.
A capillary refill time of ≥ 3 seconds is an intermediate-risk group marker for serious illness (amber sign).

Measure the BP of children with fever if the HR or capillary refill time is abnormal.

Height of body temp alone should not be used to identify children with serious illness. However, children in the following categories should be recognised as being in a high-risk:

- children < 3 months of age with a temp ≥ 38°C
- children aged 3–6 months with a temperature ≥ 39°C

Duration of fever should not be used to predict the likelihood of serious illness.

Children with fever should be assessed for signs of dehydration: prolonged capillary refill time; abnormal respiratory pattern; abnormal skin turgor; weak pulse; cool extremities.

Sxs and signs of specific illnesses

Look for a source of fever and check for the presence of sxs and signs that are associated with specific disease.

Be aware that classical signs of meningitis (neck stiffness, bulging fontanelle, high-pitched cry) are often absent in infants with bacterial meningitis.

Be aware that, in rare cases, incomplete/atypical Kawasaki disease may be diagnosed with fewer features.

Imported infections: When assessing a child with feverish illness, enquire about recent travel abroad and should consider the possibility of imported infections according to the region visited.

Table 2 Summary table for symptoms and signs of specific diseases

Diagnosis to be considered	Symptoms and signs in conjunction with fever
Meningococcal disease	Non-blanching rash, particularly with one or more of the following: an ill-looking child lesions > 2 mm in diameter (purpura) a capillary refill time of ≥ 3 seconds

	neck stiffness
Meningitis	Neck stiffness Bulging fontanelle Decreased level of consciousness Convulsive status epilepticus
Herpes simplex encephalitis	Focal neurological signs Focal seizures Decreased level of consciousness
Pneumonia	Tachypnoea (RR > 60 breaths per minute age 0–5 months, RR > 50 breaths per minute age 6–12 months; RR > 40 breaths per minute age > 12 months) Crackles Nasal flaring Chest indrawing Cyanosis Oxygen saturation ≤ 95%
Urinary tract infection	Vomiting Poor feeding Lethargy Irritability Abdominal pain or tenderness Urinary frequency or dysuria Offensive urine or haematuria
Septic arthritis	Swelling of a limb or joint Not using an extremity Non-weight bearing
Kawasaki disease	Fever for > 5 days and at least four of the following: bilateral conjunctival injection change in mucous membranes (injected pharynx, dry cracked lips or strawberry tongue) change in the extremities (oedema, erythema or desquamation) polymorphous rash cervical lymphadenopathy

Management by remote assessment

Children with any **'red'** features but who are not considered to have an immediately life-threatening illness should be urgently assessed by a healthcare professional in a face-to-face setting w/n 2 hrs.

Management by the non-paediatric practitioner

If any **'amber'** features are present and no dx has been reached, provide parents or carers with a 'safety net' or refer to specialist paeds for further assessment. Safety net should be ≥1 of the following: provide verbal and/or written info on warning sxs and how further healthcare can be accessed; arranging further FU at a specified time and place; liaising with other healthcare profs, including OOHs, to ensure direct access for the child if further assessment is required. Oral antibiotics should not be prescribed to children with fever without apparent source.

Management by the paediatric specialist

Infants < 3 mos with fever should be observed, measure and record temp, HR, and RR.

Children aged ≥ 3 months with fever without apparent source with ≥1 'red' features: perform FBC, blood culture, CRP, UA. **Consider in children with 'red' features**, as guided by the clinical assessment: LP in children of all ages (if not C/I), CXR irrespective of body temp and wcc, lytes, ABG. LP infants < 1 mo, all infants aged 1–3 mos who appear unwell and aged 1–3 mos with WBC<5 or > 15×10⁹/L.

Parenteral antibiotics should be given to: infants < 1 month; all infants aged 1–3 months who appear unwell; infants aged 1–3 months with WBC < 5 or > 15 × 10⁹/litre. When parenteral antibiotics are indicated for infants < 3 months of age, a 3rd-gen ceph (cefotaxime or ceftriaxone), should be given plus an antibiotic active against Listeria (i.e., ampicillin or amoxicillin) until culture results become available.

Children with fever without apparent source who have ≥ 1 'amber' features, perform ix's: UA; blood tests: FBC, CRP and blood cultures; consider LP for < 1 yo; CXR in a child with T > 39°C and WBC > 20 × 10⁹/litre.

Children who have been referred with fever without apparent source and no features of serious illness (**green** group), test urine for UTI and assess for sxs and signs of pneumonia. Routine blood tests and CXRs should NOT be performed with no features of serious illness (green group). Febrile children with proven respiratory syncytial virus or influenza infection assess for features of serious illness. Consider UA. In aged ≥ 3 mos with fever without apparent source, consider a period of observation in hospital (with or without ix's) to help differentiate non-serious from serious.

When a child has been given antipyretics: healthcare profs should not rely on a ↓ or lack of ↓ in temp after 1–2 hrs to differentiate between serious and non-serious illness. Children in hospital with 'amber' or 'red' features should be re-assessed after 1–2 hours.

Immediate treatment by the paediatric specialist (for children of all ages)
Children with fever and shock presenting to specialist paediatric care or A+E should be: given an immediate IV fluid bolus of 20 ml/kg. The initial fluid should be 0.9% NaCl. Actively monitor and give further fluid boluses as needed. Give immediate parenteral antibiotics if they are: shocked, unrousable, showing signs of meningococcal dis. Consider immediate parenteral antibiotics for children with fever and ↓ LOC. Seek sxs and signs of meningitis and herpes simplex encephalitis (suggestive, then give IV acyclovir). Give O_2 to children with fever who have signs of shock or SpO_2 < 92% when breathing air. Consider treatment with O_2 for children with an SpO_2 > 92%, as clinically indicated.

Treatment of suspected serious bacterial infection
In a child presenting to hospital with a fever and suspected serious bacterial infection, requiring immediate treatment, antibiotics should be directed against *Neisseria meningitidis*, *Strep pneumoniae*, *E coli*, *Staph aureus* and *Haem influenzae* type b. **A 3rd-gen cephalosporin** (i.e., cefotaxime or ceftriaxone) is appropriate, until culture results are available. For infants < 3 months of age, add an antibiotic active against Listeria (i.e., ampicillin or amoxicillin).

Admission to and discharge from hospital
In addition to the child's clinical condition, consider the following factors when deciding whether to admit:

- social and family circumstances
- other illnesses that affect the child or other family members
- parental anxiety and instinct (based on their knowledge of their child)
- contacts with other people who have serious infectious diseases
- recent travel abroad to tropical/subtropical areas, or areas with ↑ risk of endemic infectious disease
- when parent's concern for child's current illness has caused them to seek med advice repeatedly
- where the family has experienced a previous serious illness or death due to feverish illness (↑ anxiety)
- when a fever has no obvious cause, but the child remains ill longer than expected for a self-limiting illness.

Suspected meningococcal disease

Give parenteral antibiotics at the earliest opportunity (benzylpen or a 3rd-gen ceph). Admit and have their need for inotropes assessed. Give chloramphenicol if penicillin allergic.

Antipyretic interventions

Tepid sponging is NOT recommended for the treatment of fever. Children should NOT be under dressed or over wrapped. Consider the use of antipyretics in those who appear distressed or unwell. Antipyretics should not routinely be used with the sole aim of ↓ temp in children who are otherwise well. Consider the views and wishes of parents and carers. Either paracetamol or ibuprofen can be used to ↓ temp. Paracetamol and ibuprofen should NOT be administered at the same time and should not routinely be given alternately to children with fever. However, use of the alternative drug may be considered if the child does not respond to the 1st agent. Antipyretic agents do not prevent febrile convulsions and should not be used specifically for this.

Advice for parents or carers re home care of feverish child

Advise to manage their child's temp as above. Offer regular fluids (if breastfed, give breast milk).

How to detect signs of dehydration by looking for the following features: sunken fontanelle, absence of tears, dry mouth, poor overall appearance, sunken eyes. Encourage their child to drink more fluids and consider seek advice if detect signs of dehydration. How to identify a non-blanching rash; Check their child during the night

Keep their child away from nursery while the child's fever persists but to notify the school of the illness.

Following contact with a healthcare professional, seek further advice if: the child has a fit, develops a non-blanching rash, the parent or carer feels that the child is less well than when they previously sought advice, the parent or carer is more worried than when they previously sought advice, the fever lasts > 5 days, the parent or carer is distressed, or concerned that they are unable to look after their child.

PAEDIATRIC NORMAL VITAL SIGNS BY AGE GROUP
The Harriet Lane Handbook

Age group	HR	Lower limit of SBP	RR breaths/min
Birth – 6 mos	80-180 (140)	60	30-60
6 mos-1yr	70-170 (135)	70	30-50
1-3 yrs	90-150 (120)	72-86	24-40
3-5 yrs	65-135 (110)	76-80	22-34
5-12 yrs	60-120 (85-100)	80-90	16-30
12-adult	60-100 (80-85)	90	12-20

	Green – low risk Mx at home	Amber – intermediate risk Requires face to face doctor	Red – high risk 999 or see by doctor within 2hours
Colour	Normal colour of skin, lips and tongue	Pallor reported by parent/carer	Pale/mottled/ashen/ blue
Activity	Responds normally to social cues Content/smiles Stays awake or awakens quickly Strong normal cry/not crying	Not responding normally to social cues Wakes only with prolonged stimulation Decreased activity No smile	No response to social cues. Appears ill to a healthcare professional Unable to rouse or does not stay awake Weak, high-pitched or continuous cry
Respiration		Nasal flaring Tachypnoea: RR > 50 breaths/minute age 6–12 months RR > 40 breaths /minute age > 12 months Oxygen saturation ≤ 95% in air Crackles	Grunting Tachypnoea: 60 breaths/minute Moderate or severe chest indrawing
Hydration	Normal skin and eyes Moist mucous membranes	Dry mucous membrane Poor feeding in infants CRT ≥ 3 seconds Reduced urine output	Reduced skin turgor

Other •	**None** of the amber or red symptoms or signs	Fever for ≥ 5 days	Age 0–3 months, temperature ≥ 38°C
			Age 3–6 months, temperature ≥ 39°C
		Swelling of a limb or joint	Non-blanching rash
			Bulging fontanelle
		Non-weight bearing/not using an extremity	Neck stiffness
			Status epilepticus
			Focal neurological signs
			Focal seizures
		A new lump > 2 cm	Bile-stained vomiting

CRT, capillary refill time; RR, respiratory rate.

CONGENITAL HEART DISEASES

Atrial septal defect precordial bulge, prominent pulmonary artery pulsations, mitral regurgitation murmur, fixed splitting of S-2. L-R ` shunting, embolic stroke.

Coarctation of the aorta prominent brachial pulses with weak or nonpalpable femoral arterial pulses; systolic murmur in the L infraclavicular area and under the L scapula.

Tetralogy of Fallot R ventricular outflow tract obstruction; VSD; Overriding aorta; RVH. Male. Cyan.

Ventricular septal defect defect in interventricular septum, harsh systolic murmur at lower L sternal border, laterally displaced apical impulse, recurrent URTIs, acyanotic.

CHOLESTEROL

The BHS and JBS2 guidelines have stated that the **ideal cholesterol targets** are: to lower **TC** by 25% or **LDL** cholesterol by 30% or to reach < **4.0 mmol/l or** < **2.0** mmol/l respectively, whichever is the greater - however a TC < 5.0 mmol/l or LDL cholesterol < 3.0 mmol/l or reductions of 25% or 30%, respectively (whichever is the greater), provides a minimal acceptable "audit" standard. Also CVD risk replaces CHD risk estimation to reflect the importance of stroke prevention as well as CHD prevention. The new CVD risk threshold of ≥ 20% is equivalent to a 10 yr CHD risk of approx ≥ 15%.

Who should receive statins? DTB 2001

Statins provide preventative benefit for patients with clinically obvious atherosclerotic disease, CHD, and for those without features but with an absolute risk of ≥ 15% or more over 10 years. However, treating all those with a 10 yr CHD risk of ≥ 15% would in effect include around 25% of UK adults and is not achievable with current NHS funding. A more realistic approach, therefore, is to ensure that all those with an absolute 10-yr CHD risk > 30% receive optimum statin tx, with appropriate monitoring of lipid profile and advice on non-drug measures, and extending statin tx to remaining individuals with a risk level of at least 15% over 10 years, as resources permit. Special consideration needs to be given to: diabetics whose risk of developing and dying from CHD is 2-5 x that of non-diabetics); those of South Asian descent (40% ↑ risk of developing CHD than white UK pops); and patients with familial hypercholesterolaemia (who are at particularly high risk of dying from CHD). Families of patients with inherited dyslipidaemia, i.e. familial hypercholesterolaemia, including children, deserve similar attention.

Foods That Raise Triglyceride Levels and should be limited.

Alcohol. Saturated fats: animal fats, lard, butter and shortening, fried foods, whole milk dairy products, cheese, cream cheese, high-fat meats and fast foods. **Trans fats:** Hydrogenated fats in margarine, vegetable shortening, fried foods, fast foods, pastries, cakes, pies, crackers. **Sugar:** Concentrated sweets (sugar, honey, molasses, jams, jellies and candy). Desserts. **Drinks:** Fruit juices, fruit drinks, fruit punches, sodas, smoothies, sports drink and sweetened coffee. **Others:** Sweetened cereals, flavored yogurts and sports or energy bars. **Starch:** Bagels, pasta, rice, potatoes, large rolls, pizza, pretzels, popcorn, chips, many fat-free foods and ready-to-eat cereals. Choose **small portions** due to their **high carb density**. Use whole grains and legumes (starchy beans) in preference to refined starches.

BEST Food Choices: Fresh Fruit: Avoid fruit juices even fresh squeezed (high sugar content). **Veg:** Eat at least $\geq$ 3 cups a day of fresh or frozen prepared veg. **Breads and cereals:** Choose whole grain breads, crackers, unsweetened high-fiber cereals or old fashioned oatmeal. Try amaranth, millet, quinoa, barley, buckwheat or bulgur. **Protein:** Lean meats, poultry without skin, egg, egg substitute or white, cooked dried beans, lentils, peas, nuts, low-fat soy. **Eat fish** minimum 2x a wk. Include fatty fish (salmon, mackerel, blue fin tuna, sardines, anchovies). **Dairy:** Fat-free or 1% milk, fat-free or lowfat plain yogurt, light fruit yogurt, low-fat or soy cheeses. **Fat:** Include 1 tablespoon of canola, olive or peanut oil/ day. Include 1/4 cup of nuts/day (almonds, walnuts, peanuts, mixed nuts) or 1/2 avocado. Avocados, olives and natural nut butters.
Sugar-free   products: Diet   drinks,   gelatin   and   pudding.

CHROMOSOMAL ABNORMALITIES
(The Merck Manual of Diagnosis and Therapy)
AUTOSOMAL

Downs Syndrome (Trisomy 21; Trisomy G; Mongolism)
- Incidence 1/800, mothers > 40 yo = 1/40 incidence; > 20 yo = 1/2000. Mean IQ 50
- 95% cases trisomy 21; 5% translocation (t) = t(14;21) extra 21 on 14 or t(21;22) or mosaicism (one normal cell and one with 47 chromsomes)
- Placcid newborn, muscular hypotonia, extra neck skin fold (oedema), microcephaly, flat occiput, short stature, epicanthal folds, flat bridge of nose, large tongue lacking central fissure, ears small
- Brushfield's spots (gray to white spots around edge of iris)
- Short, broad, hands, single simian palmar crease, clinodactyly (incurved 5th finger with 2 phalanges)
- Wide feet with gap between 1st and 2nd toes, plantar furrow, dermatoglyphics (characteristic dermal prints of hands and feet). 40% congenital heart disease (ventricular septum or AV canal), duodenal atresia; hearing, visual, thyroid
- Females have 50% chance of having child with Down's. Men are infertile. Death 5th or 6th decade.

Trisomy 18 (Edwards Syndrome; Trisomy E)
- 1/6000 live births. Extra 18

- SGA, hypotonia, hypoplasia of skeletal muscle and fat, weak cry; polyhydramnios, small placenta, single umbilical artery; hypoplastic orbital ridges, short palpebral fissures, small mouth and jaw
- Microcephaly, prominent occiput, low-set malformed ears, short sternum
- Clenched fist with index finger overlapping 3rd and 4th fingers. Absent distal crease on 5th finger, hypoplastic fingernails. Dorsiflexed short big toe, clubfeet and rocker-bottom feet common
- Congenital heart disease, anomalies of lungs, diaphragm, abdominal wall, kidneys, ureters
- Hernias+/or diastasis recti, cryptorchidism, redundant posterior neck skin folds
- < 10% alive at 1 year. Marked developmental delay and disability.

Trisomy 13 (Patau Syndrome; Trisomy D)
- 1/10,000 live births, extra 13
- Midline abnormalities, holoprosencephaly (failure of forebrain to divide properly), cleft lip, cleft palate, microphthalmia, colobomas (fissures) of iris, retinal dysplasia. Shallow supraorbital ridges, slanted palpebral fissures, low-set ears, deafness. SGA. Simian crease, polydactyly, hyperconvex narrow fingernails. 80% congenital heart disease. Dextrocardia. Scalp defects, dermal sinuses. Loose skin posterior neck. Cryptoorchidism, abnormal scrotum. Bicornuate uterus. Severe mental retardation. < 10 % survive > 1 year.

DELETION SYNDROMES
5p-Deletion (cri du chat syndrome)
- Deletion of short arm of 5. High-pitched cry for several weeks only. LBW. Microcephaly. Round face with wide-set eyes. Strabismus. Low set ears. Narrow EAC, preauricular tags. Heart defects. Hypotonic infants. Mental and physical retardation, Disabled if survive to adulthood.
4p-Deletion (Wolf-Hirschhorn syndrome)
- Severe MR. Broad or beaked nose. Midline scalp defects. Colobomas and ptosis, cleft palate. Hypospadias and cryptorchidism. Severely handicapped if into 20s. Epilepsy
Contiguous gene syndromes (microdeletions)
Telomeric deletions (deletions at end of chromosome)

SEX CHROMOSOME
Turner Syndrome (Bonnevie-Ulrich Syndrome)
- 1/4000 live female births. 50% 45XO rest mosaics (45,X/46,XX or 45,X/47,XXX)
- Dorsal lymphedema of hands and feet, loose skin folds posterior neck. Short stature, neck webbing, low hairline on back of neck, ptosis. Broad chest with wide-spaced nipples, multiple pigmented naevi.
- Short 4th metacarpals and metatarsals, finger pads with whorl demographics. Nail hypoplasia, increased carrying angle at elbows. Aortic coarc, bicuspid aortic valve. Renal anomalies and haemangiomas. GIT telangiectasias – GIB. Rare to have mental deficiency. Gonadal dysgenesis (no breast tissue, menses); HRT to bring on puberty. Rarely are fertile. Risk of gonadoblastoma (CA).

Triple X Syndrome (47, XXX) - phenotype female
1/1000 of normal females have this. Sterility, menstrual irregularity, IQ < 90.

Klinefelter Syndrome (47, XXY) - phenotype male, male hypogonadism
1/800 live male births. Infertile, tall, long arms and legs, small firm testes, 1/3 gynaecomastia, delayed puberty, loss of libido, reduced facial hair, osteoporosis (brittle bones), depression, early heart disease. Many have deficits in verbal IQ, processing and reading. Rx - testosterone replacement, cosmetic surgery.

47, XYY Syndrome
1/1000 males. Tall. Hyperactive, ADD, learning disabilities, minor behavioural disorders.

INTERSEX STATES
Female pseudohermaphrodites norm internal genitalia + ovaries but ambiguous ext. Aberration on 6.
Male pseudohermaphrodites testicular gonadal tissue. 46 XY. Ambiguous external genitalia.
Female phenotype seen in the complete form of testicular feminization synd.
True hermaphrodites. Both ovarian and testicular tissue. Mixed masculine and feminine genital structures. Most 46, XX karyotype.
Mixed gonadal dysgenesis (short, 46, XY/45,XO, ambiguous genitalia, short)
Pure gonadal dysgenesis (bilateral streaks - primitive gonadal tissues).

INHERITANCE

Cystic fibrosis has a simple Mendelian autosomal recessive inheritance. Affected individuals will have 2 copies of the mutated CFTR gene, one inherited from each parent. Carriers will have one normal and one mutated CFTR gene and their health will not be affected. However carriers will have the potential to pass on the gene to their offspring. Brothers and sisters of affected individuals are at ↑ risk (1 in 4) of having CF because both parents will be carriers.

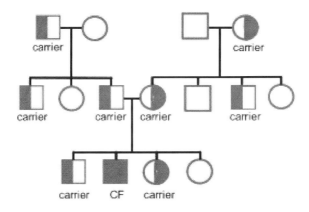

Autosomal recessive inheritance

If 2 carriers of the mutated CF gene have children then there is: a 1 in 4 chance that their baby will have CF, a 1 in 4 chance that their baby will not have CF or carry a CFTR mutated gene and a 2 in 4 chance that the baby will not have CF but will carry 1 CFTR mutated gene. In cystic fibrosis, the lack of CFTR function caused reduced fluid production and enhanced sodium absorption through epithelial Na+ channels (ENaC) and basolateral Na/K ATPase pumps. This results in ↑ fluid absorption leading to drier airways and impaired ciliary clearance.

Delta F508 gene occurs in 75% of cystic fibrosis patients in the UK. A three base-pair deletion in exon 10 results in the omission of phenylalanine at position 508 of CFTR, leading to a combination of defective intracellular processing (which results in an absence of CFTR from the membrane) and defective channel function. http://www.cysticfibrosismedicine.com.

Autosomal recessive: Alpers Syndrome (neurodegenerative disease) , Autosomal Recessive Polycystic Kidney Disease, Cystic Fibrosis, Hereditary haemochromatosis, Sickle Cell Anemia, Spinal Muscular Atrophy type I, Spinal Muscular Atrophy type II, Spinal Muscular Atrophy type III, Tay Sachs , Thalassemia, Usher Syndrome.

X-linked recessive: Becker muscular dystrophy, deuteranopia, Duchenne muscular dystrophy, fragile X syndrome, haemophilia

Autosomal dominant: achondroplasia, acute intermittent porphyria, Charcot Marie Tooth (CMTIA), Huntington's disease, Marfan's syndrome, myotonic dystrophy, NF1 (chromosome 17 – von Recklinghausen's disease).

<u>CHRONIC FATIGUE SYNDROME (NICE Aug 2007)</u>

Consider CFS/ME if a pt has: Fatigue with ALL of the following features: new or had a specific onset (not lifelong), persistent and/or recurrent, unexplained by other conditions has resulted in a substantial ↓ in activity level, characterised by post-exertional malaise and/or fatigue (typically delayed, by at least 24 hours, with slow recovery over several days) and ≥ 1 of the following sxs:

- difficulty with sleeping (insomnia, hypersomnia, unrefreshing sleep, a disturbed sleep–wake cycle)
- muscle and/or joint pain that is multi-site and without evidence of inflammation
- headaches
- painful lymph nodes without pathological enlargement
- sore throat
- cognitive dysfunction (difficulty thinking, inability to concentrate, impairment of short-term memory, and difficulties with word-finding, planning/organising thoughts and info processing)
- physical or mental exertion makes symptoms worse
- general malaise or 'flu-like' symptoms
- dizziness and/or nausea
- palpitations in the absence of identified cardiac pathology.

Be aware that the sxs of CFS/ME fluctuate in severity and may change in nature over time.

Signs and sxs that can be caused by other serious conditions ('red flags') should not be attributed to CFS/ME without consideration of alternative dxs or comorbidities. The following features should be investigated:

- localising/focal neurological signs
- signs and symptoms of inflammatory arthritis or connective tissue disease
- signs and symptoms of cardiorespiratory disease
- significant weight loss
- sleep apnoea
- clinically significant lymphadenopathy.

<u>Hx</u> (exacerbating and alleviating factors, sleep disturbance and intercurrent stressors) and physical and assessment of psychological wellbeing should be

carried out. A child who has sxs suggestive of CFS/ME should be referred to paeds for assessment to r/o other dx w/n 6 wks of presentation.

Ixs: UA for protein, blood and glucose; FBC; urea and lytes; LFTs: TFTs; ESR or plasma viscosity; CRP; random blood glucose; serum creatinine; screening blood tests for gluten sensitivity, serum calcium; creatinine kinase assessment of serum ferritin levels (children and young people only).

Serum ferritin in adults should not be carried out unless a FBC and other haematological indices suggest Fe deficiency.

Vitamin B_{12} deficiency and folate levels should not be carried out unless FBC and MCV show a macrocytosis.

Serological testing should not be carried out unless the hx is indicative of an infection. Depending on the hx, tests for the following infections may be appropriate:

- chronic bacterial infections, i.e. borreliosis
- chronic viral infections, i.e. HIV or hepatitis B or C (BBC news: dx of hepatitis C was missed in a pt with CFS!)
- acute viral infections, i.e. infectious mononucleosis (use heterophile antibody tests)
- latent infections, i.e. toxoplasmosis, Epstein–Barr virus or cytomegalovirus.

Advice on sx mx should not be delayed until a diagnosis is established. This advice should be tailored to the specific symptoms the person has, and be aimed at minimising their impact on daily life and activities.

If sxs do not resolve as expected in a pt initially suspected of having a self-limiting condition, listen carefully to the pt's and their family and/or carers' concerns and be prepared to reassess their initial opinion.

Consider discussion with a specialist if there is uncertainty.

A dx is made after other possible dxs have been excluded and the sxs have persisted for: 4 months in an adult or 3 months in a child or young person; the dx should be made or confirmed by a paediatrician.

Realistic info about CFS/ME and encourage cautious optimism. Most will improve over time and some will recover and be able to resume work and normal activities. However, others will continue to experience sxs or relapse and some with severe CFS/ME may remain housebound. The prognosis in children and young people is more optimistic.

The dx of CFS/ME should be reconsidered if none of the following key features are present:
- post-exertional fatigue or malaise
- cognitive difficulties
- sleep disturbance
- chronic pain.

Sx mx- There is no known drug tx or cure for CFS/ME. However, sxs should be managed as in usual clinical practice. If patients with CFS/ME have concerns, consider starting drug tx for CFS/ME symptoms at a lower dose than in usual clinical practice. The dose may be increased gradually, in agreement with the pt.

Although exclusion diets are not generally recommended for managing CFS/ME, many find them helpful in managing sxs, including bowel sxs. If undertaking an exclusion diet or dietary manipulation, seek advice from a dietitian because of the risk of malnutrition.

Provide tailored sleep management advice that includes: Explaining the role and effect of disordered sleep or sleep dysfunction in CFS/ME. Identifying the common changes in sleep patterns seen in CFS/ME that may exacerbate fatigue sxs (insomnia, hypersomnia, sleep reversal, altered sleep–wake cycle and non-refreshing sleep). Providing general advice on good sleep hygiene. Introducing changes to sleep patterns gradually.

If sleep mx strategies do not improve the person's sleep and rest, the possibility of an underlying sleep disorder or dysfunction should be considered, and interventions provided if needed.

Sleep mx strategies should not include encouraging daytime sleeping and naps.

Rest periods are a component of all mx strategies for CFS/ME. Advise people with CFS/ME on the role of rest, how to introduce rest periods into their daily routine, and the frequency and length appropriate for each pt. Limit the length of rest periods to 30 minutes at a time. Introduce 'low level' physical and cognitive activities (depending on the severity of sxs).

Relaxation techniques should be offered for the mx of pain, sleep problems and comorbid stress or anxiety. There are a number of different relaxation techniques (guided visualisation or breathing techniques) that can be incorporated into rest periods. Pacing helpful in self-managing CFS/ME. Advise that, at present, there is insufficient research evidence on the benefits or harm of pacing. Well-balanced diet in line with 'The balance of good health'. They should work to develop strategies to minimise complications that may be caused by nausea, swallowing problems, sore throat or difficulties with buying, preparing and eating food. Eating regularly, and including slow-release starchy foods in meals and snacks. For moderate or severe CFS/ME, providing or recommending equipment and adaptations (wheelchair, blue badge or stairlift) should be considered as part of an overall mx plan, taking into account the risks and benefits for the individual. This may help them to maintain their independence and improve their quality of life.

Having to stop their work or education is generally detrimental to patient's health and well-being. Therefore, the ability to continue in education or work should be addressed early and reviewed regularly. Proactively advise about fitness for work and education, and recommend flexible adjustments or adaptations to work or studies to return to them when they are ready and fit enough. This may include, with the informed consent of the pt, liaising with employers, education providers and support services, such as: OH services, disability services through Jobcentre Plus, schools, home education services, local ed authorities, disability advisers in universities and colleges.

For patients who are able to continue in or return to education or employment, ensure, with the patient's informed consent, that employers, occupational health or education institutions have info on the condition and mx plan.

Referral to specialist CFS/ME care should be offered: w/n 6 months of presentation to patients with mild CFS/ME
- within 3–4 months of presentation to patients with moderate CFS/ME symptoms
- immediately to patients with severe CFS/ME symptoms.

Specialist CFS/ME care After a patient is referred, an initial assessment should be done to confirm the diagnosis.
If general mx strategies are helpful, these should be continued after referral to specialist CFS/ME care.

CBT, graded exercise therapy and activity mx programmes. Prescribing of low-dose TCAs, specifically amitriptyline, should be considered for people with CFS/ME who have poor sleep or pain. Should not be offered to people who are already taking SSRIs because of the potential for serious adverse interactions. Melatonin may be considered for children and young people with CFS/ME who have sleep difficulties, but only under specialist supervision because it is not licensed in the UK. Advise that setbacks/relapses are to be expected as part of CFS/ME.

CHRONIC HEART FAILURE
(NICE July 2003, updated Aug 2010)

Diagnosis

- Hx, sxs (breathlessness, fatigue), signs (fluid retention). 12-lead ECG +/or BNP or NTproBNP
- Other recommended tests: cxr, u/e, creatinine, FBC, TFTs, LFTs, FBG, lipids, UA, PFR or spirometry
- If ECG or BNP abnormal, arrange transthoracic Doppler 2-D echo to exclude valve disease, assess systolic + diastolic function of the LV and detect intracardiac shunts. If the echo is normal, consider diastolic dysfunction, refer. If abnormal echo, assess severity, aetiology, precipitating factors, type of cardiac dysfunction, consider refer.
- Measure BNP or NTproBNP in suspected CHF without prior MI (new).
- If BNP > 400 or NTproBNP > 2000 pg/ml or if previous MI with suspected CHF, arrange transthoracic Doppler 2-D echo and specialist assessment within 2 weeks (new 2010).

Treatment

- Aerobic exercise training/ Refer to smoking cessation services/ Abstain from EtOH if EtOH-related.
- Annual flu jab; Pneumococcal vaccine x 1/ Air travel depends on clinical condition at time of travel/ DVLA.

Drug treatment for LV systolic dysfunction

- Offer ACEI + β-blockers to all patients with LVSD. Use clinical judgement to decide which first (new 2010).
- Offer β-blocker to all patients with CHF due to LVSD including older adults and patients with PVD, erectile dysfunction, DM, COPD without reversibility and interstitial pulmonary disease (new 2010).
- Generalist: Start ACEI and titrate upwards q 2/52. Check urea, creat and lytes after initiation and at each dose increment. Or if not tolerated (severe cough), consider AIIR antagonists. Add diuretic to control

congestive sxs and fluid retention. Diuretic is 1st line if patient presents with acute pulmonary oedema.

- Specialist: Add β-blocker and titrate upwards after diuretic + ACEI treatment. Start low, go slow manner with assessment of HR, BP and clinical status after each titration. Add digoxin if patient in sinus rhythm remains symptomatic despite tx with diuretic, ACEI + β-blocker OR if patient is in AF then use as 1st-line. Avoid triple tx of ACEI, β-blocker + AIIR antagonist pending results of further trials.
- Add spironolactone (12.5-50mg od) if patient remains mod to severely symptomatic despite optimal drug tx. Check K, creat for signs of ↑K (1/2 dose of spironolactone and recheck) and renal failure.
- Amiodarone - made by specialist. Need 6-monthly clinical review, LFTs, TFTs, side-effects.
- Anticoagulants - heart failure + AF; heart failure + h/o thromboembolism, LV aneurysm or intracardiac thrombus
- Aspirin (75-150 mg od) for patients with heart failure + atherosclerotic arterial disease (including coronary heart disease)
- Statins for patients with ht F + atherosclerotic vascular disease in accordance to current indications
- Isosorbide/ hydralazine combo who are intolerant of ACEI or ATIIR antagonist (specialist only)
- IV inotropic agents (dobutamine, milrinone or enoximone) short-term tx for acute decompensation of chronic heart failure
- Calcium channel blockers - amlodipine considered for treatment of co-morbid HTN +/or angina in patients with heart failure. Avoid verapamil, diltiazam or short-acting dihydropyridine agents.

Invasive Procedures

- Coronary revascularisation - refractory angina
- Cardiac transplantation - severe refractory sxs or refractory cardiogenic shock
- Cardiac resynchronisation tx - LVEF<=35%, drug refractory sxs, and a QRS duration > 120 ms.
- Implantable cardioverter-defibrillators (ICDs) - cardiac arrest due to VT of VF, spontaneous sustained VT causing syncope, sustained VT without syncope/ cardiac arrest and who have EF < 35% but are no worse than Class III NYHA; h/o prior MI + non-sustained VT on

Holter, inducible VT on electrophysiological testing, LVEF < 35% and no worse than Class III; familial cardiac condition with high risk of sudden death (long QT syndrome, hypertrophic cardiomyopathy, Brugada syndrome, arrhythmogenic RV dysplasia and following repair of Tet of Fallot).

Clinical review q 6months

- Hx, NYHA class/ PE - weight, JV distension, lung crackles, hepatomegaly, peripheral oedema, lying + standing BP
- 12-lead ECG or 24h Holter if suspicion of arrhythmia.
- Urea, lytes, creat (and eGFR – new 2010). TFTs, FBC, LFT and level of anticoagulation may be required depending on rx. No routine digoxin monitoring. A digoxin concentration w/n 8-12h of last dose to confirm clinical impression of toxicity or non-compliance.

Referral for specialist advice

- Seek specialist advice and consider adding one of the following if a patient remains symptomatic despite optimal tx on ACEI and a β-blocker: aldosterone antagonist licensed for heart failure (especially if NYHA Class III-IV) or has had an MI within the past month or an ARB (if NYHA Class II-III) or hydralazine in combination with nitrate (if Afro-Carribean and has NYHA Class III-IV) (new 2010).
- Heart failure due to valve disease (no ACEI), diastolic dysfunction (dx + tx by specialist. Treat with low to med dose of loop diuretics < 80 mg frusemide /day. If not respond, refer), or any other cause except LV systolic dysfunction.
- ≥ 1 co-morbidities (COAD/asthma/ reversible airways disease; renal dysfunction (creatinine > 200 umol/L), anaemia, thyroid disease, PVD, urinary frequency, gout)
- Angina, atrial fibrillation (Specialist to decide whether to improve HR control or cardiovert (return to sinus).
- Anticoagulation is indicated or other symptomatic arrhythmia. Severe heart failure/ heart failure not responding to treatment as discussed in guideline/ heart failure that cannot be managed in the home. Women who are planning a pregnancy or who are pregnant.
- When a patient is admitted to hospital with heart failure, seek advice from a specialist in heart failure (new 2010).

**UK Guidelines for Identification, Mx and Referral for
CHRONIC KIDNEY DISEASE in adults
(Sources: RCGP Introducing eGFR, promoting good CKD management;
www.renal.org/ CKD guide)**

The KDOQI stages of chronic kidney disease are:

Stage	GFR	Description	Treatment stage
1	90+	Normal kidney function but urine or other abnormalities (i.e. known to have proteinuria, haematuria (no urological cause), microalbuminuria (DM), polycystic disease or reflux nephropathy)	Observation, control of BP, eGFR urine PCR if dipstick protein present; yearly
2	60-89	Mildly reduced kidney function, urine or other abnormalities point to kidney disease	BP control, monitor as per stage 1. No further testing if eGFR alone.
3	30-59	Moderately reduced kidney function	also Hb, K, phosphate, Ca 6 monthly (12 if stable < 2 mL /min change eGFR over 6/12)
4	15-29	Severely reduced kidney function	Also bicarb, PTH. Plan for RRR. 3 monthly (6 if stable CKD stage 4) Refer urgently.
5	≤ 14	Very severe, or **endstage** kidney failure (sometimes call **established renal failure**)	As per stage 4. Refer urgently.

eGFR 100% kidney function = 100 ml/min based on creat, gender, age (↓ with age). Multiply by 1.2 for Afro-Carribean. Does not apply to ARF or age < 18. Underestimates severity in muscle wasting or an amputation.

**CKD if eGFR < 60 ml/min/1.73m2 or the presence of kidney damage for >
3/12**: review all previous creat/ eGFR results to assess rate of deterioration. Review meds, (NSAIDs, antibiotics, mesalazine, diuretics, ACEIs/ ARBs.

Test urine for haematuria and proteinuria. If proteinuria, request urine ACR.

Assess clinically: for urinary symptoms, palpable bladder, BP, sepsis, heart failure, hypovolaemia. Repeat creatinine within 5 days to r/o rapid progression if new finding. Check referral criteria: ensure entry into a chronic disease management programme if not indicated.

Causes of CKD: hypertension +DM (#1), GN (haematuria, proteinuria, red cell casts), congenital, PKD (loin pain, fhx, x-ray), reflux nephropathy (tubulointerstitial disease, h/o uti, reflux). **Sxs**: uraemia: fatigue, SOB, wt loss, N/V, loss of appetite, oedema. Everyone normally loses 0.75 ml/yr after 45.

CKD leads to: hypertension, fluid retention, ↑K, ↑PO4, metabolic acidosis, (activation of RAAS –LVH), anaemia (hb ↓ with ↓eGFR = fatigue), lack of production of vitamin 1,25 D (osteitis fibrosa, osteomalacia) leads to PO4 retention, low Ca and then hyperparathyroidism, coronary heart disease

Ix for CKD: creat, urea, K, HCO3, Ca, PO4, PTH, Hb; urine microscopy + protein; U/S, CT, nuclear scan; renal biopsy (glom).

Mx for CKD: BP and cholesterol control, inhibition of RAS, ↓proteinuria, stop smoking, control BG.

Tx for CKD: salt restriction, low K and PO4 diet, NaHCO3 for met acidosis; epoitin 6000U/wk and iron (IV if ferritin < 200) so target hb 10.5-12.5 (check Hb 2wkly) and ferritin 200-500 ng/ml; vitamin D2 supplements and Ca-based PO4 binders, if high PTH start 1α calcidol, if high Ca, low PTH, then stop 1α calcidol; ACEI BP control < 130/80.

Renal replacement tx: if eGFR < 10, sxs of uraemia and malnutrition, intractable volume overload, persistent ↑K or met acid. Choices: home or hospital haemodialysis 3x a week for 4h via fistula, graft or catheters; peritoneal dialysis (continuous ambulatory or automated); kidney transplant (cadaveric, live or preemptive), immunosuppression (IL2 antibody, mycophenolate mofetil, steroids, tacrolismus). Complications: CA, infection, rejection.
Example: 58 yo NIDDM with hb 10, Ca 2.4, PO4 1.6, PTH 20 (1.5-6.5), protein 900 mg/day, creat 300 μmol/l, BP 140/85.
Mx: start ramipril and also 1α calcidol for epo deficiency.

Referral Criteria:

Stage 1/2: Urgent: malignant hypertension; K > 7 mmol/l; nephrotic syndrome

Routine: isolated proteinuria (protein: creat ratio PCR > 100 mg/mmol); proteinuria and microscopic haematuria (PCR > 45 mg/mmol); diabetes with proteinuria (PCR > 100 mg/mmol) but no retinopathy; macroscopic haematuria (after negative urological evaluation); recurrent pulmonary oedema with normal LV function; fall of eGFR of > 15% during 1st 2/12 on ACEI/ ARB.

Stage 3 As above plus: progressive fall in GFR; microscopic haematuria (after negative urological tests if > 50 yo); proteinuria (urine PCR > 45 mg/mmol); anaemia (after exclusion of other causes); persistently abnormal K, Ca, PO4 (uncuffed sample); suspected SLE, vasculitis, myeloma; uncontrolled hypertension BP > 150/90 on 3 drugs.

Stage 4/5 Urgent. All patients should be referred and offered the options of renal replacement therapy (RRT) or conservative tx, even if RRT will not be appropriate. Exceptions may include if the CKD is part of terminal illness or function is stable and relevant tests completed and appropriate mx implemented with agreed tx plan.

Info needed on referral: gen med hx, urinary sxs, rxs, exam (BP, oedema, bladder), urine dipstick for blood and protein, urine for PCR if proteinuria present, FBC, creat, urea, Na, K alb, Ca, PO4, cholesterol, HbA1C (DM), all previous creat results with dates, results of renal scan if available.

Mx all stages: Regular clinical and lab assessment. Advice on smoking, weight, exercise, salt and EtOH intake. CV prophylaxis : if risk > 20% at 10 yrs, consider aspirin if BP < 150/90 and lipid lowering rxs (or entry into trials).

- **Meticulous BP control.** Threshold 140/90, target 130/80 in most patients; threshold 130/80, target 125/75 if urine PCR > 100 mg/mmol. Include ACEI or ARB if urine PCR > 100 mg/mmol or if diabetes and microalbuminuria present. Check creat and K before starting and 2 wks after start and after each dose change. If creat ↑ by > 20% or GFR ↓ by > 15% repeat with K and seek advice (? Stop ? test for RAS renal artery stenosis)

- If K > 6 mmol/L, check no haemolysis and check diet; stop NSAIDs and LoSalt (K containing salt substitute); stop K retaining diuretics; stop ACEI/ARB if hyperkaemia persists.
  CKD Stage 3: additional mx to include:
- If Hb < 11 and other causes excluded, refer for IV iron +/- ESA (erythropoietin stimulating agent) with target Hb 11-12 g/dl.
- Renal scan if lower urinary tract sxs, refractory hypertension, unexpected falling eGFR.
- Immunise against influenza and pneumococcus.
- Review all drugs ensure correct dose; avoid nephrotoxic drugs (NSAIDs) if possible.
- Check PTH level when stage 3 first diagnosed. If high, check 25-hydroxy vitamin D and if low, give ergo- or cole-calciferol with Ca supplement (not PO4), repeat PTH after 3/12 and refer if still high.

CKD Stage 4/5: additional mx in conjunction with 2° care:
- Dietary assessment and info about tx options
- Immunize against hepatitis B
- Mx of hyperparathyroidism
- Correction of acidosis
- Timely dialysis access procedure
- Referral/ discussion even if dialysis may not be appropriate

DRUGS TO AVOID IN CKD AND DRUGS WHICH ARE SAFE

Antihypertensive agents: Thiazide diuretics not recommended if serum creat > 2.5 mg/dL or creat clearance < 30 mL/min. Loop diuretics are most common drugs for uncomplicated hypertension in CKD. K-sparing diuretics and aldosterone blockers can ↑ serum K. ACE inhibitors and angiotensin receptor blockers (ARBs) are 1st line for patients with diabetes mellitus and early kidney dis. ACE inhibitors and ARBs can cause GFR ↓ and creat ↑, especially if congestive heart failure, diuretic or NSAID use, or at high dose; discontinuation recommended if serum creat ↑ > 30% or serum K at least 5.6 mEq/L. No adjustment needed for metoprolol tartrate, metoprolol succinate, propranolol, labetalol, calcium channel blockers, clonidine, and α-blockers.

Hypoglycemic agents: Metformin ↑ risk for lactic acidosis and not recommended if serum creat >1.5 mg/dL in men or 1.4 mg/dL in women, age > 80 years, or there is chronic heart failure. Sulfonylureas can cause severe hypoglycemia and should not be used in stages 3 to 5 CKD. No adjustment needed for glipizide.

Antimicrobial agents: Pen G or carbenicillin can cause neuromuscular toxicity, myoclonus, seizures, or coma.
Imipenem/cilastatin can cause seizures. Tetracyclines, except doxycycline, can exacerbate uremia.
Nitrofurantoin metabolite can cause peripheral neuritis. Aminoglycosides should not be used if possible.

Analgesic agents: Meperidine, dextropropoxyphene, morphine, tramadol, and codeine metabolites can affect CNS and respiratory systems and are not recommended in stage 4 or 5 CKD. No adjustment needed for paracetamol.

NSAIDs: Use is linked to 3x higher risk for acute renal failure. Use can cause nephrotic syndrome with interstitial nephritis and chronic renal failure. ↓ K excretion can lead to hyperkalemia. ↓ sodium excretion can lead to peripheral oedema, ↑ BP, and exacerbation of heart failure. Antihypertensive effects of β-blockers, ACE inhibitors, or ARBs can be decreased. Cyclooxygenase 2 inhibitors have similar renal effect. Short-term NSAID use is well tolerated if pt is well hydrated and has good renal function and absence of heart failure, diabetes, or hypertension. Long-term use not recommended. Check serum creatinine every 2 to 4 weeks in early tx.

Statins: Dosing adjustment is recommended except for atorvastatin.
Herbal products: St. John's wort and ginkgo can ↑ metabolism of other medications. Ginkgo ↑ bleeding risk if taking aspirin, ibuprofen, or warfarin. Alfalfa, dandelion, and noni juice contain potassium. Products with heavy metals and Chinese herbal products with aristolochic acid are nephrotoxic. Vasoconstrictive ingredients can cause hypertension.

Renal Unit of Royal Infirmary of Edinburgh
Role of kidney: to remove toxic waste products and excess H20 & salts; control BP; produce Epo to stimulate red cell production from BM (risk of anaemia); keeps Ca and PO4 in balance for healthy bones; maintains blood in neutral non acid state.

Fluid and salt problems May have to follow fluid restriction or take diuretics (frusemide), and to restrict salt intake to prevent fluid overload (swollen ankles or SOB). Give a 'target' or 'ideal' weight (dry weight). Occasionally some have the opposite problem, and need to take extra fluid and salt.

BP Meticulous control has been shown to slow down the progression of CKD. ACEI (or similar) will be recommended, particularly good for kidney disease.

Anaemia is mainly due to a deficiency of a hormone **Epo.** Often the anaemia of renal failure can be helped by taking iron. Some people remain short of iron even when taking iron tablets. If so, might need a course of IV iron injections as a hosp out pt. With more severe anaemia prescribe Epo for self-admin as injections, usually 1x or 2x a wk.

Bones Renal bone disease can be a serious problem for those who have had CRF for a long time, causing aches, pains and sometimes fractures. The aim is to prevent it, as treatment is much harder later. In renal failure typically the Ca level in the blood becomes low and the PO4 level high. This imbalance needs tx, or the body overproduces PTH in an attempt to control it, and this causes thinning of the bones. May have to take a combination of the following: **Alfacalcidol or calcitriol**: these are active forms of Vitamin D, which is often short in renal failure, as the kidneys fail to process it. **Phosphate binders** (eg Phosex, Calcichew, Calcium 500, Renagel) help to prevent too much absorption of phosphate from the gut and thus keep the level in the blood lower. **Diet** may also help with this. Follow a special diet (dietitian) in which there may be controlled amounts of protein, salt, phosphate, and K. May not need to restrict all of these, and recommendations are likely to change with time. Avoid under-feeding.

Acidaemia In renal failure the kidneys are unable to excrete the normal acid waste products of the body. In renal failure often have too much acid in the blood (acidaemia) and have to take bicarb tablets to neutralize.

Prevention of heart disease, stroke and vascular disease ↑ risk. Stop smoking, maintain healthy diet and take regular exercise. Cholesterol and other lipids are often high in kidney disease, this may require rx.

Stages 3-4, each visit assess fluid balance, BP, diet and nutrition, anaemia, prevention of bone disease, prevention of heart disease, stroke and vascular disease.

COPD MANAGEMENT
NICE Guidelines June 2010

Diagnosis: Assess lung function (Spirometry $FEV_1/FVC < 0.7$ or if $FEV_1 > 80\%$, then with cough or breathlessness).

MRC dyspnoea score.

Grade	Degree of breathlessness related to activity
1	Not troubled by breathlessness except on strenuous activity.
2	SOB when walking or hurrying up a slight hill
3	Walks slower than others on level ground or has to stop for breath when walking at own pace.
4	Stops for breath after walking on level ground 100 m or after a few minutes
5	Too breathless to leave the house, or when dressing/undressing.

Investigations: CXR; Hb to exclude anaemia and polycythaemia; BMI.

Additional: CT chest; ECG; Echo; Pulse oximetry; Sputum culture

Serial PFMs showing 20% diurnal variation suggests asthma; Transfer factor for carbon monoxide; Alpha-1-antitrypsin deficiency if early onset or + FH

Mild or moderate – assess annually; Severe – assess twice yearly

Classification Stage 1 Mild ($\geq 80\%$ predicted), Stage 2 Mod (50-80% predicted FEV_1), Stage 3 Severe (30-49%), Stage 4 Very Severe ($< 30\%$).

Consider diagnosis: > 35 yo; Smoker/ ex-smoker; No signs of asthma

Symptoms (exertional breathlessness, chronic cough, regular sputum production, frequent winter bronchitis or wheeze)

Definitions: Long-acting bronchodilator (LABA); inhaled corticosteroid (ICS); long-acting muscarinic antagonist (LAMA).

COPD Management Algorithm

Breathless and Exercise Limitation (SABA or SAMA prn)

$\downarrow$ $\downarrow$

Exacerbations or persistent breathlessness

$FEV_1 \geq 50\%$ $\qquad\qquad\qquad\qquad$ $FEV_1 < 50\%$

$\downarrow$ $\qquad\qquad\qquad\qquad\qquad$ $\downarrow$

LABA or LAMA (stop SABA) $\quad$ LABA + ICS or LAMA (stop SABA)

(Combo inhaler)

$\downarrow$ $\qquad$ $\downarrow$ $\qquad\qquad\qquad$ $\downarrow$ $\qquad$ $\downarrow$

Persistent exacerbations or breathlessness

$\downarrow$ $\qquad\qquad$ $\downarrow$ $\qquad\qquad\qquad\qquad$ $\downarrow$

LABA + ICS $\quad\rightarrow\quad$ LAMA + (LABA + ICS in combo inhaler)

Mx of COPD exacerbations: ↑ bronchodilator, neb, antibiotics if sputum, prednisolone 30 mg 7-14d or admit.

Theophylline only to be used after a trial of SABA and LABA or unable to use inhaled therapy.

Additional supportive measures

Encourage stop smoking (NRT, Varenicline, or bupropion). Pulmonary rehab if recent hospitalisation. Mucolytic therapy for chronic productive cough. Non-invasive ventilation for persistent hypercapnic ventilatory failure.

Long-term O_2 therapy for PaO_2 < 7.3 kPa or < 8 kPa with FEV 1 < 30%, O_2 sats < 92%, with 2° sxs (cyanosis, polycythaemia, ↑ JVP, peripheral oedema) for at **least 15h a day, 20 h bettet**. Ambulatory and short burst O_2 should be given as needed. Address obesity and poor nutrition. Flu/ pneumoccocal vaccine. Identify and treat depression associated with COPD. Self-mx advice for exacerbation. GPs should only prescribe **large 1360L cylinder** for those in palliative care, for existing patients while awaiting respiratory assessment and post assessment not requiring O_2 and unwilling to stop using (if ≤ 2 cylinders/wk) consultant will advise post review, during early supported discharge until post discharge assessment re continuation or stopping.

Specialist referral

Diagnosis uncertainty
Bullous lung disease
Frequent infections - r/o bronchiectasis
Dysfunctional breathing
Onset of cor pulmonale
Assessment for pulmonary rehab
Rapid decline in FEV1
< 40 yo or FH of α-1-antitrypsin deficiency
Symptoms disproportionate to lung function deficit
Haemoptysis - r/o bronchial CA
Assessment for O_2 therapy, long-term nebuliser or oral steroid therapies
Severe COPD
Assessment for surgical options – lung transplantation, lung volume reduction surgery

CONSULTATION MODELS

Roger Neighbour's 'The Inner Consultation,' 1987. 5 checklist stages are:

Connecting – building + establishing rapport with the pt. May include skills i.e. acceptance set, curtain raiser/ opening gambit, internal search, matching, NLP, and speech censoring.

Summarising – getting to the point of why the pt has come, using skills of eliciting to discover their ICE and summarising back to the patient. The listening and eliciting skills include the pt is right to start with, explain why you are asking, be facilitative and encouraging (with open-ended questions, statements), echoing and checking.

Handing over – doctors' and patients' agenda agreed. Include negotiating, influencing (doctor's apostolic function) and gift-wrapping. **Safety-netting** – 'What if?' predicting skills - what would the doctor do in each case.

Housekeeping – Take care of yourself - Am I in good enough shape for the next patient? Get up, stretch, coffee.

1984 Pendleton D. 'The Consultation, An Approach to Learning and Teaching'

The 7 tasks to an ideal consultation are: 1. To define the reason for the patient's attendance, including: the nature and history of the problem, their cause, the patient's ICE, the effects of their problems 2. To consider other problems: at risk factors, continuing problems. 3. To choose with the pt an appropriate action for each problem.4. To achieve a shared understanding of the problem with the patient. 5. To involve the patient in the mx plan and encourage him to accept appropriate responsibility 6. To use time and resources appropriately in the consultation in the long-term 7. To establish or maintain a relationship with the patient which helps to achieve the other tasks.

1981 Helman's Folk Model
– Cecil Helman is a medical anthropologist and suggests a patient with a problem comes to a doctor seeking answers to 6 questions: What has happened? Why has this happened? Why to me? Why now? What would happen if nothing was done about it? What should I do about or whom should I consult for further help?

1979 Stott and Davis

A	B
Management of presenting problems	Modification of help seeking behaviours

C	D
Management of continuing problems	Opportunistic health promotion

1976 Byrne and Long 'Doctors Talking to Patient' A study of 2500 audio-taped consultations led to a description of 6 phases, which occur in a consultation: 1. The doctor establishes a relationship with the patient. 2. The doctor either attempts to discover or actually discovers the reason for the patient's attendance.3. The doctor conducts a verbal or physical exam or both. 4. The doctor, or the doctor and the patient, or the patient considers the condition.5. The doctor and occasionally the patient details tx or further ix.6. The consultation is terminated usually by the doctor.

1975 John Heron's Six Category Intervention Analysis - Heron is a humanist psychologist who described the behaviour of health professionals or the interventions of a doctor as 1 of 6 interventions:

prescriptive – give advice or instructions, being critical or directive

information – impart new knowledge, instructing or interpreting

confrontational – challenge a restrictive attitude or behaviour, giving direct feedback within a caring context

cathartic – seek to release emotion in the form of weeping, laughter, trembling or anger

catalytic – encourage the patient to discover and explore his own latent thoughts and feelings

supportive – offer comfort and approval, affirming the patient's intrinsic value

1966 Eric Berne's 'Games People Play' Transactional Analysis – he classifies the states of mind as parent, adult and child and that an individual has a given repertoire of behaviour corresponding to this state of mind.

The Child – spontaneous or dependent. Many GP consultations are conducted between a parental doctor and a child-like patient and may not be in the best interests of the patient. The doctor should be aware of transactional analysis and be flexible enough to change his repertoire to avoid consultations degenerating into the games people play.

The Adult – this logical ego state is concerned with problem solving, taking in data and processing, and storing knowledge and skills.

The Parent – critical or caring, nurturing or controlling. The transactions during consultations may be complimentary, crossed or ulterior.

1957 Balint, M. 'The Doctor, His Patient and the Illness'

The 3 main themes are: psychological problems are often manifested physically and physical disease causes psychological problems, doctors have feelings too and these feelings can impact on the consultation process, and doctors can be trained in a limited way to be more sensitive to what is going on in the patient's mind.

Balintian concepts: The Dr as a Drug. The Child as the Presenting Complaint – ticket of entry. Elimination by Appropriate Physical Exam. Collusion of Anonymity – no-one taking final responsibility for the patient. The Flash – the real reason for attendance is made apparent to both the doctor and the patient. The Mutual Investment Company – the patient presents with episodic offers of both physical and psychological problems in a long relationship.

CONTRACEPTION - Missed Pills

> 12h late	≤ 12h late
Take the last pill you missed.	Take the last pill you missed now.
Leave any missed pills.	
Use condoms for next 7 days.	
7 or more pills left in pack after the missed pill	< 7 pills left
When you have finished the pack, leave the usual 7 day break before you start the next pack	When you have finished the pack, start the next pack and omit the 7 day break.

If you start a new pack late, you may need **Emergency Contraception**. EC is recommended:
- if ≥ 2 20 mcg or 3 30-35 mcg ethinylestradiol tablets are missed from the 1st 7 tablets in a pack and if had UPSI in pill-free wk or wk 1
- if ≥ 4 consecutive tablets are missed mid-packet (need to omit pill free break with new pack)

EC - < 72h after UPSI = Levonelle - 2 (two 750 mcg levonorgestrel pill STAT, i.e. 1.5g)

EC < 5 days after UPSI = IUCD

EC > 5 days but up to 5 days after the earliest likely calculated ovulation = IUCD

Use **condoms (additional contraceptive precaution) for 7 days during and for 7 days after** if you:
- Vomit up to 2-3h after taking the pill or have severe diarrhoea
- As of March 2011, no need to use additional precautions with antibiotics.
- Taking drugs, which induce hepatic enzyme activity:

phenytoin, phenobarbital, carbamazepine, nelfinavir, primidone
rifampicin, griseofulvin, nevipariner, rifabutin, topiramate
modafinil, oxycarbazepine , ritonavir

* Continue condoms for 7 days after stopping this drug. For rifabutin and
rifampicin, additional contraceptive precaution required for 4 weeks after
stopping a short-term course.

**The United Kingdom Medical Eligibility Criteria for Contraceptive Use
(UKMEC***)* recommends that breastfeeding F should not use COC (UKMEC
Category 4, unacceptable health risk) in the first 6 weeks postpartum.
Between 6 weeks and 6 months postpartum, risks of COC use for
breastfeeding F outweigh the benefits (UKMEC Category 3). After 6 months
postpartum, the benefits of COC use generally outweigh the risks.

Contraception after abortion: you may start coc on same day as TOP. It is
immediately effective as long as the coc is started within 6 days of the TOP.

Postnatal contraception: needed by day 21 (ovulate by day 28). 4 wks
(IUS/IUD). Anytime in non-breast feeding F on pop, implant, injectable. 6/12
COC and 6 wks injectable in breast-feeding F. 6/52 for caps and diaphragms.

UKMEC Changes 2009 BMI > 40 removed as UKMEC4 (do not use). CVA
UKMEC3 (with expert judgement) for IUS continuation. Age restriction for
migraine without aura has been removed. No migraine with aura for 5 years is
UKMEC3 for COC. Fibroids are UKMEC 3. Liver disease and SLE are new
categories.

Nexplanon replaced Implanon Sept 2010. X-ray can locate barium in
nexplanon. No longer need to scan to locate.
IUS used for 5 yrs or up to 7 yrs if fitted from age 45. TT380s has more
copper so is more effective and lasts 10 yrs.

EllaOne: new EC for up to 120h; inhibits or delays ovulation. 85%
pregnancies prevented 0-72h vs. 69% for L-2.

Peri-menopause: Stop contraception 2 yrs after LMP < 50yo, 1 yr after LMP
> 50yo, assume infertile at 55, stop coc and injection at 50 yo; POP & implant,
if no periods > 50 yo, measure FSH and continue for 1 yr if raised.

Serious health events caused or prevented by HRT, per 1,000 women (estrogen 0.625 mg plus progestin 2.5 mg). WHI 2002

* Risk is greatest during the first 2 years of use.

** First signs appear during the first year of use.

*** Risk first appears after 4 years of use.

**** Risk first appears after 1 year of use.

\# Benefit appears after 3 years of use

Health event	After 2 years of HRT use	After 5.2 years of HRT use
Blood clots (venous thromboembolism)	6 more*	9 more
Coronary artery disease	3 more**	4 more
Breast cancer	No change***	4 more
Stroke	1 more****	4 more
Colorectal cancer	No change	3 fewer#
Hip fractures	1 fewer	2 fewer
Death	No change	No change

Million Women Study - 14 cases breast CA/ 1000 non-HRT users. Extra breast CA cases in 1000 HRT users for 5 yrs HRT use is 1.5 (RR=1.3) (oestrogen only) and 6 (RR=2) (combined HRT). *Lancet* 2003; 362:419-427.

WHI (women's health initiative) trial and HERS (heart and estrogen/progestin replacement study RCT of 2,763 postmenopausal women in US with IHD) showed small ↑ in risk of CHD in 1st year of use of combined HRT and no cardioprotection even in oestrogen-only arm of WHI or after tx for 4.1 yrs in HERS. *JAMA* 2004;291:1701-1712. WHI trial found ↑ risk of stroke in HRT users. ↑risk of DVT in 1st year of use.

Dec 2003: CSM published advice - HRT should still be considered for short-term use for menopausal sxs because the benefits still outweigh the risks - lowest effective dose for the shortest duration.

CONTROLLED DRUGS

The Misuse of Drugs Regs 2001 divides CDs into 5 schedules corresponding to their therapeutic usefulness and misuse potential. A number of changes affecting the prescribing, record keeping and destruction of CDs have been introduced as a result of amendments to the Misuse of Drugs Regs 2001. The Controlled Drugs (Supervision of Management and Use) Regs 2006 came into effect 1st Jan 2007.

Schedule 1 (CD licence) Have no recognised medicinal use and include hallucinogenic drugs (coca leaf, LSD and mescaline). Production, possession and supply of these drugs are limited to research or other special purposes. Practitioners and pharmacists may not lawfully possess Schedule 1 drugs except under licence.

Schedule 2 (CD) Includes the opiates (diamorphine, heroin, morphine, pethidine), secobarbital, glutethimide, amphetamine and cocaine.
- Are subject to safe custody requirements (except for secobarbital) and so must be stored in a locked receptacle, an appropriate CD cabinet or approved safe, which can only be opened by the person in lawful possession of the CD or a person authorised by that person.
- A licence is required to import or export drugs in Schedule 2.
- The drug may be administered to a patient by a doctor or dentist, or by any person acting in accordance with the directions of a doctor or dentist.
- A register must be kept for Schedule 2 CDs and this register must comply with the relevant regs.
- The destruction of CDs in Schedule 2 must be appropriately authorised and the person witnessing the destruction must be authorised to do so.

Schedule 3 (CD No Register)
- Includes a small number of minor stimulant drugs and other drugs which are less likely to be misused than the drugs in Schedule 2. Barbiturates (except secobarbital, now Schedule 2).
- Are exempt from safe custody requirements and can be stored on the open dispensary shelf except for flunitrazepam, temazepam, buprenorphine and diethylpropion, which must be stored in a locked CD receptacle.

- Are subject to the same special handwriting requirements as Schedule 2 CDs, except for temazepam and phenobarbital. Phenobarbital and temazepam can be dispensed in response to a computer-generated prescription but the prescriber's signature must be added by hand.
- There is no legal requirement to record transactions in a CD register.
- The requirements relating to destruction do not apply unless the CDs are manufactured by the individual. Retain invoices for a minimum of 2 years.

Schedule 4 (CD Benzodiazepines and CD Anabolic steroids)
- Are exempt from safe custody requirements, with destruction requirements only applying to importers, exporters and manufacturers. Specific CD prescription-writing requirements do not apply.
- CD registers do not need to be kept, although records should be kept if such CDs are produced, or if a licensed person imports or exports such drugs.
- Part 1 (CD Benzodiazepines): Includes most of the benzos, including zolpidem, plus 8 other substances including fencamfamin and mesocarb. Except temazepam (which is now Schedule 3).
- Possession of is an offence without an appropriate rx. Possession by GPs and pharmacists acting in their professional capacities is authorised.
- Are subject to full import and export control.
- Part 2 (CD Anabolic steroids): Includes most of the anabolic and androgenic steroids (testosterone), clenbuterol (adrenoreceptor stimulant), somatotropin, somatropin, somatrem HCG, and non-HCG.
- There is no restriction on the possession when it is part of a medicinal product.
- A Home Office licence is required for the importation and exportation of substances unless the substance is in the form of a medicinal product and is for self-administration by a person.

Schedule 5 (CD Invoice): Includes certain CDs (codeine, pholcodine, morphine) which are exempt from full control when present in medicinal products of low strengths as their risk of misuse is reduced.
- No restriction on the import, export, possession, administration or destruction of these preparations and safe custody regulations do not apply.
- A practitioner, pharmacist or a person holding an appropriate licence

may manufacture or compound any CD in Schedule 5. Invoices must be kept for a minimum of 2 years.

Prescriptions for Controlled Drugs

The amendments to the Misuse of Drugs Regulations 2001 that came into force in Nov 2005 removed the requirement that CD prescriptions should be written in the prescriber's own handwriting was removed. This means that CD prescriptions can be type-written, handwritten or computer printed. Only the signature of the prescriber has to be handwritten. Further changes following amendments to the Misuse of Drugs Regulations came into force in July 2006.

- A new requirement that patients or other people collecting medicines on their behalf must sign for them.
- Validity of any rx for schedule 2, 3 & 4 controlled drugs to be restricted to 28 days.
- Strong recommendation that the max quantity is limited to 30 days for rxs of schedule 2, 3 and 4.
- Re-emphasis of professional guidance that doctors should prescribe controlled drugs for themselves or family members only in exceptional circumstances.
- Drs are only able to prescribe diamorphine, dipipanone and cocaine to substance misusers for the treatment of addiction if they hold a licence issued by the Home Office. All doctors may prescribe such drugs for patients, including substance misusers, for the relief of pain due to organic disease or injury without a specific licence.
- Rx's for temazepam and for Schedule 4 and 5 controlled drugs are exempt from the specific prescription requirements of the Misuse of Drugs Regulations 2001. However, they must still comply with the general prescription requirements as specified under the Medicines Act.
- Emergency supplies of Schedule 2 and 3 controlled drugs, for a specific patient, are not permitted either at the request of the patient or a practitioner, except for phenobarbital for the tx of epilepsy.

Keeping and Storage of Schedule 2 Controlled Drugs and Buprenorphine

- These CDs must be kept in a locked receptacle. This can be a doctor's bag with a lock, and if the bag is transported in the dry's car, it must be locked and placed in a locked boot. A locked car is not adequate; the bag must also be locked.
- In the surgery, a locked cabinet should ideally be secured to the wall of a room and only opened by the GP or someone authorised by the GP.

- Records for Schedule 2, but not Schedules 3, 4 or 5, CDs must be kept in a CD register. All health care professionals (GPs, pharmacists, etc.) must keep their own CD register and keep it up to date.

Doctor's Bag

- Where a doctor carries a bag for home visits containing CDs, a separate CD register should be kept for the CD stock held within that bag. Each doctor is responsible for the receipt and supply of CDs from their bag.
- Restocking of a bag for home visits, etc. from practice stock should be witnessed by another member of the practice staff, as should the appropriate entries into the practice's CD register.
- Where a rx is written by a doctor following the administration of a CD to a patient, the doctor should endorse the rx form with the word 'administered' and then date it. Enter info into the patient's record as soon as is practicable.

Prescribing in Instalments

- Some CDs can be dispensed in instalments providing they are prescribed using specific NHS prescription forms.
- In England, GPs must use the form FP10MDA-S to prescribe in instalments Schedule 2 CDs, **buprenorphine (Schedule 3)** or diazepam (Schedule 4) for drug addiction. This form must not be used for any other purpose.
- The rx must be dispensed on the date on which it is due. If the client does not collect an instalment when it is due that supply is no longer valid. The client cannot collect that supply the following day.
- If a controlled drug prescription is to be dispensed in instalments, e.g. daily, then the FP10MDA-S must specify: The number of instalments. The intervals to be observed between instalments (if necessary, instructions for supplies at weekends or bank holidays should be included). The total quantity of controlled drug that will provide treatment for a period not exceeding 14 days. The quantity to be supplied in each instalment.
- Current legislation does not allow Schedule 2 and 3 controlled drugs to be prescribed as repeat rx's.

CORONER (need to inform)

Sudden or unexpected deaths; accidents and injuries; industrial disease (mesothelioma); service disability pensioners; deaths where doctor has not attended w/n the past 14 days; deaths arising from ill-treatment (abuse, hypothermia, neglect, starvation); cause of death unknown; deaths < 24h after hospital admission; poisoning (chronic alcoholism and its sequela no longer per se); medical mishaps (anaesthetics; short-term, long-term operation complications, drugs therapeutic or addictive); abortions; prisoners; stillbirths (when there is doubt as to whether the baby was born alive).

Rights of deceased: Patient's trustee or surviving spouse has rights to deceased patient's records and to decision-making.

CREMATION FORMS (Ministry of Justice www.justice.gov.uk 2008)

CR1: Application for cremation by the executor. **CR2**: Application for cremation of body parts.

CR3: Application for cremation of stillborn baby.

CR4: Medical certificate filled by doctor who has treated the deceased during the last illness and has normally seen the deceased w/n 14 days of death. Requires 2 independent doctors to complete 4 and 5 to corroborate the medical circumstances in which death took place. It is normal practice for doctors from different hospital units to sign 4 and 5 or to arrange a local GP to sign 5. If a junior hospital doctor signs 4, then the doctor in charge may not sign 5. A GP partner may sign 4, if he has seen the deceased outside the normally acceptable period (14d) and the attending partner is unavailable.

CR5: Confirmatory Medical Cert filled by a registered medical practitioner of ≥ 5 yrs' standing, who is not a relative or work colleague of the deceased or a relative or partner of the doctor who has signed Form 4.

COMPLETION OF DEATH CERTIFICATES

http://www.sehd.scot.nhs.uk/cmo/CMO(2009)10.pdf

If you do not know the cause of death, refer the death to the Procurator Fiscal.

Old age should only be given as a cause of death if: you have personally cared for the deceased over a long period (yrs or many mos), you have observed a gradual decline in your patient's general health and functioning, you are not aware of any identifiable disease or injury that contributed to death, you are certain that there is no reason that the death should be reported to the PF, the patient is ≥ 80. Never use 'natural causes' alone.

I (a) Pathological fractures of femoral neck and thoracic vertebra I (b) severe osteoporosis I (c) old age II fibrosing alveolitis

I (a) Old age II Non-insulin dependent diabetes mellitus, essential hypertension and diverticular disease

I (a) hypostatic pneumonia I (b) Dementia I (c) Old age

Organ failure I (a) Congestive cardiac failure I (b) Essential hypertension

I (a) Renal failure I (b) Necrotising proliferative nephropathy I (c) SLE II Raynaud's phenomenon and vasculitis

I (a) Liver failure I (b) Hepatocellular cancer or Liver cirrhosis I (c) Chronic hep B infection and alcoholism (joint)

I (a) Cardiorespiratory failure I (b) Ischaemic heart disease and Chronic obstructive airways disease II osteoarthritis

CVD I (a) Subarachnoid haemorrhage I (b) Ruptured aneurysm of anterior communicating artery

I (a) Intraventricular haemorrhage I (b) Warfarin anticoagulation I (c) Atrial fibrillation I (d) Ischaemic heart disease

I (a) Cerebral infarction I (b) Thrombosis of basilar artery I (c) Cerebrovascular atherosclerosis I (d) Hypertension

Neoplasms: I (a) Carcinomatosis I(b) Small cell carcinoma of left main bronchus I(c) Heavy smoker for 40 yrs

II Hypertension, Cerebral arteriosclerosis, Ischaemic heart disease

I (a) Intraperitnoeal haemorrhage I (b) Ruptured metastatic deposit in liver I (c) From primary adenocarcinoma of ascending colon II NIDDM (no abbrev)

I (a) Post-transplant lymphoma I (b) Immunosuppression I (c) Renal transplant I (d) Glomerulonephrosis due to IDDM

I (a) Pathological fractures of left shoulder, spine and shaft of right femur I (b) Widespread skeletal secondaries I (c) From Primary Adenocarcinoma of breast II Hypercalcaemia

I (a) Lung metastases I (b) From testicular teratoma

I (a) Massive haemoptysis I (b) Primary small cell carcinoma of left main bronchus II Primary adenocarcinoma of prostate

I (a) Multiple organ failure I (b) Poorly differentiated metastases throughout abdominal cavity I (c) Unknown primary site

Leukemia: I (a) Neutropenic sepsis I (b) Acute myeloid leukemia
I (a) Haemorrhagic gastritis I (b) Chronic lymphatic leukemia II Myocardial
ischaemia, Valvular heart disease

Diabetes: I (a) End stage renal failure I (b) Diabetic nephropathy I (c)
Insulin dependent diabetes mellitus, Type 2
I (a) Septicaemia – fully sensitive Staphylococcus aureus I (b) Gangrene of
both feet due to PVD I (c) NIDDM, Type 2

Infections: I (a) Bilateral pneumothoraces I (b) Multiple bronchopulmonary
fistulae I (c) Extensive cavitating pulmonary tuberculosis (smear and culture
positive) II Iron-deficiency anaemia, ventilator associated pseudomonas
pneumonia
I (a) Bronchopneumonia I (b) Immobility and wasting I (c) Alzheimer's
disease
I (a) Lobar pneumococcal pneumonia I (b) Influenza A II Ischaemic heart
disease; I (a) Meningococcal septicaemia

Injury: I (a) Pulmonary embolism I (b) Hemiarthroplasty 2 days after
Fractured neck of femur I (c)Tripped on loose floor rug at home II Left-sided
weakness and difficulty with balance since haemorrhagic stroke 5 years ago

DEPRESSION (NICE OCTOBER 2009)

Guideline Development Group decided to adopt DSM-IV criteria not ICD-10.
Symptoms should be present for at least 2 weeks and each symptom should be present at sufficient severity for most of every day. Both diagnostic systems require at least 1 (DSM-IV) or 2 (ICD-10) key symptoms (low mood, loss of interest and pleasure or loss of energy) to be present. Increasingly, depressive sxs below the DSM-IV and ICD-10 threshold criteria can be distressing and disabling if persistent. Therefore this updated guideline covers 'subthreshold depressive symptoms', which fall below the criteria for major depression, and are defined as at least one key symptom of depression but with insufficient other sxs and/or functional impairment to meet the criteria for full diagnosis. Symptoms are considered persistent if they continue despite active monitoring and/or low-intensity intervention, or have been present for a considerable time, typically several months. (For a dx of dysthymia, sxs should be present for at least 2 years.)

DSM-IV Diagnosis
Subthreshold depressive symptoms: < 5 symptoms of depression.
Mild depression: few if any symptoms in excess of 5 required to make the diagnosis and symptoms result in only minor functional impairment
Mod depression: symptoms or functional impairment between mild and severe
Severe depression: most symptoms and the symptoms markedly interfere with functioning. Can occur with or without psychotic symptoms.

Principles for assessment Conduct a comprehensive assessment that does not rely simply on a symptom count. Take into account the degree of functional impairment and/or disability associated with the possible depression & the duration of the episode.
Consider how the following factors may have affected the development, course and severity of a person's depression:
any history of depression and comorbid mental health or physical disorders, any past history of mood elevation (to determine if the depression may be part of bipolar disorder); any past experience of, and response to, treatments; the quality of interpersonal relationships; living conditions and social isolation.

Step 1: Case identification and recognition Be alert to possible depression (particularly in patients with a past hx of depression or a chronic physical health problem with assoc functional impairment) and consider asking 2 questions:

1. During the last month, have you often been bothered by feeling down, depressed or hopeless?
2. During the last month, have you often been bothered by having little interest or pleasure in doing things?

Step 2: For persistent subthreshold depressive symptoms or mild to moderate depression

Advice on sleep hygiene: establishing regular sleep and wake times, avoid excess eating, smoking or drinking alcohol before sleep, create a proper environment for sleep, taking regular physical exercise.

Low-intensity psychosocial interventions consider offering ≥ 1 of the following interventions: individual guided self-help based on the principles of cognitive behavioural therapy (CBT), computerised cognitive behavioural therapy (CCBT), a structured group physical activity programme.

Step 3: persistent subthreshold depressive symptoms or mild to moderate depression with inadequate response to initial interventions, and moderate and severe depression: an antidepressant (normally a SSRI]) **or** a high-intensity psychological intervention option: CBT, interpersonal therapy (IPT), behavioural activation (but note that the evidence is less robust than for CBT or IPT) , behavioural couples therapy for people who have a regular partner and where the relationship may contribute to the development or maintenance of depression, or where involving the partner is considered to be of potential therapeutic benefit.

For moderate or severe depression, provide a combo of antidepressants & high-intensity psychological intervention (CBT or IPT).

For patients with depression who decline an antidepressant, CBT, IPT, behavioural activation and behavioural couples therapy, consider: counseling for patients with persistent subthreshold depressive sxs or mild to moderate depression, short-term psychodynamic psychotherapy for patients with mild to moderate depression.

Drug tx SSRIs are associated with ↑ risk of bleeding, especially in older patients or those taking other rxs that have the potential to damage the GI mucosa or interfere with clotting. Consider a gastroprotective drug in older patients taking NSAIDs or aspirin.

Fluoxetine, fluvoxamine and paroxetine are associated with a higher propensity for drug interactions than other SSRIs.

Paroxetine is associated with a higher incidence of discontinuation symptoms than other SSRIs.

Take into account toxicity in overdose when choosing an antidepressant for people at significant risk of suicide. Be aware that: compared with other equally effective antidepressants recommended for routine use in primary care, venlafaxine is associated with a greater risk of death from overdose

TCAs, except for lofepramine, are associated with the greatest risk in overdose.

When prescribing rxs other than SSRIs, take the following into account: The increased likelihood of the patient stopping tx because of side effects (and the consequent need to increase the dose gradually) with venlafaxine, duloxetine and TCAs.

The potential for higher doses of venlafaxine to exacerbate cardiac arrhythmias and the need to monitor BP.

The possible exacerbation of hypertension with venlafaxine and duloxetine

The potential for postural hypotension and arrhythmias with TCAs

The need for haematological monitoring with mianserin in elderly people. Dosulepin should not be prescribed.

Non-reversble MAOIs, such as phenelzine, should normally be prescribed only by specialist mental health pros.

For people started on antidepressants who are not considered to be at increased risk of suicide, review after 2 weeks. See them regularly at intervals of 2 to 4 weeks in the first 3/12, and then at longer intervals if response is good.

A patient with depression started on antidepressants who is considered to present an ↑ suicide risk or is < 30 yrs (because of the potential ↑ prevalence of suicidal thoughts in the early stages of antidepressants for this group) should normally be seen after 1 week and frequently thereafter until the risk is no longer considered clinically important.

If a patient with depression develops side effects early in antidepressants, provide appropriate info and consider one of the following strategies: monitor sxs closely where side effects are mild and acceptable to the person **or** stop the rx or change to a different antidepressant if the person prefers **or** in discussion with the person, consider short-term concomitant tx with a benzodiazepine if anxiety, agitation and/or insomnia are problematic (except in patients with chronic sxs of anxiety); this should usually be for no longer than 2 weeks in order to prevent the development of dependence.

If response is absent or minimal after 3 to 4 wks of tx with a therapeutic dose of an antidepressant, ↑ the level of support (i.e., by weekly face-to-face or telephone contact) and consider: ↑ the dose in line with the SPC if there are no significant side effects **or** switching to another antidepressant if there are side effects or if the person prefers.

When switching to another antidepressant, which can normally be achieved within 1 week when switching from drugs with a short half-life, consider the potential for interactions in determining the choice of new drug and the nature and duration of the transition. Exercise particular caution when switching: from fluoxetine to other antidepressants, because fluoxetine has a long half-life (1 week). When stopping an antidepressant, gradually reduce the dose, normally over a 4-week period, although some people may require longer periods, particularly with drugs with a shorter $t_{1/2}$ (such as paroxetine and venlafaxine). This is not required with fluoxetine because of its long $t_{1/2}$.

Do not use antidepressants routinely to treat persistent subthreshold depressive sxs or mild depression because the risk–benefit ratio is poor, but consider them for people with: a past hx of moderate or severe depression or initial presentation of subthreshold depressive sxs that have been present for at least 2 years or subthreshold depressive sxs or mild depression that persist(s) after other interventions.

Continuation and relapse prevention Support and encourage a patient who has benefited from taking an antidepressant to continue for at least 6 months after remission of an episode of depression. Discuss that this greatly reduces the risk of relapse and antidepressants are not associated with addiction.

Psychological interventions for relapse prevention Patients with depression who are considered to be at significant risk of relapse (including those who have relapsed despite antidepressants or who are unable or choose not to continue antidepressant tx) or who have residual symptoms, should be offered one of the following psychological interventions: individual CBT for those who have relapsed despite antidepressant rx and for patients with a significant hx of depression and residual sxs despite tx; mindfulness-based cognitive therapy for people who are currently well but have experienced ≥ 3 previous episodes of depression.

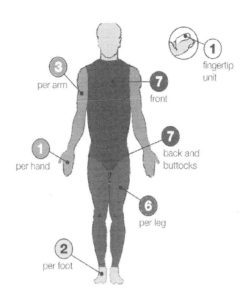

DERMATOLOGY
Know your finger-tip unit!

Steroid Potency – National Psoriasis Foundation (www.psoriasis.org.uk)/ British National Formulary

BRAND NAME	GENERIC NAME
CLASS 1 – Superpotent (Very Potent)	
Clobex 0.05%; Cormax 0.05%; Olux E Foam, 0.05%; Temovate 0.05%, Dermovate 0.05%, Clarelux, Etrivex	Clobetasol propionate
Diprolene Ointment, 0.05%	Betamethasone dipropionate
Ultravate Cream/Ointment, 0.05%	Halobetasol propionate
Vanos Cream, 0.1%	Fluocinonide
Psorcon ointment, 0.05%; Psorcon E ointment, 0.05%	Diflorasone diacetate
Nerisone forte	Diflucortilone valerate

CLASS 2 – Potent (palms and soles of feet) not > 40-50g/week

Diprolene Cream AF, 0.05%	Betamethasone dipropionate
Elocon Ointment, 0.1%	Mometasone furoate
Florone Ointment, 0.05%; Psorcon Cream 0.05%	Diflorasone diacetate
Halog Ointment/Cream, 0.1%	Halcinonide
Lidex Cream/Gel/Ointment, 0.05%	Fluocinonide
Topicort Cream, 0.25%; Topicort Gel, 0.05%	Desoximetasone

CLASS 3 - Upper Mid-Strength (Potent)

Cutivate Ointment, 0.005%	Fluticasone propionate
Lidex-E Cream, 0.05%	Fluocinonide
Luxiq Foam, Betnovate, Betacap, Bettamousse, Diprosone	Betamethasone valerate (esters)
Betnovate C, Betnovate N, Fucibet, Lotriderm	Betamethasone esters with antimicrobials
Topicort LP Cream, 0.05%	Desoximetasone

CLASS 4 - Mid-Strength (torso, extremities) not > 90-100g/ week

Cordran Ointment, 0.05%	Flurandrenolide
Elocon Cream, 0.1%	Mometasone furoate
Kenalog , 0.1%, Triadcortyl (with antimicrobial), Aureocort (with antimicrobial)	Triamcinolone acetonide
Synalar Ointment, 0.03%, Synalar C with antimicrobial, Synalar N with antimicrobial	Fluocinolone acetonide

| Westcort Ointment, 0.2% | Hydrocortisone valerate |

CLASS 5 - Lower Mid-Strength (Moderate)

Eumovate, Trimovate with antimicrobial	Clobetasone butyrate
Betnovate RD, 0.025%	Betamethasone valerate
Capex Shampoo, 0.01%; Synalar Cream,	Fluocinolone acetonide
Cordran Cream/Lotion/Tape, 0.05%	Flurandrenolide
Cutivate Cream/Lotion, 0.05%	Fluticasone propionate
DermAtop Cream, 0.1%	Prednicarbate
DesOwen Lotion, 0.05%	Desonide
Locoid, 0.1%; Pandel Cream, 0.1%	Hydrocortisone
Alphaderm, Calmurid HC	Hydrocortisone with urea
Westcort Cream, 0.2%	Hydrocortisone valerate

CLASS 6 - Mild

Aclovate Cream/Ointment, 0.05%	Alclometasone dipropionate
Derma-Smoothe/FS Oil, 0.01%; Synalar Cream/Solution, 0.01% (1 in 10 dilution)	Fluocinolone acetonide
Desonate gel, 0.05%; Verdeso Foam, 0.05%	Desonide

CLASS 7 - Least Potent (face, genitalia)

Cetacort Lotion, 0.5%/1%; Cortaid Cream/Spray/Ointment, 1%; Hytone Cream/Lotion, 1%/2.5%; Micort-HC Cream, 2%/2.5%; Nutracort Lotion, 1%/2.5%; Synacort Cream, 1%/2.5%; canesten HC; Daktacort; Dioderm; Efcortelan; Timodine; Vioform-H	Hydrocortisone

Contact dermatitis itchy form of eczema; hairdressers to dye products; nickel jewellery, children to synthetic clothing (polyester, rayon, acrylic, acetate)

Enterobius itchy anus due to threadworms. Common in children. Treat with oral mebendazole. May need to repeat tx in 2/52 when the eggs hatch. Children ingest from fingernails.

Lichen planus itchy, shiny, violaceous papules with an overlying linear white streaks; Koebner phenomenon; seen in molluscum contagiosum, plane warts, psoriasis, vitiligo.

Pediculus humanus apply topical aqueous cabaryl 1% liquid or 0.5% lotion, malathion.

Pediculosis capitis (head lice) 0.5% liquid or lotion or permethrin 1% cream rinse for 12 hours or overnight and again 7 days to prevent lice emerging from any surviving egg.

Sarcoptes scabiei intractable pruritis at night to waste products of mites as they burrow in skin. Burrows seen in fingerwebs, flexor aspect of wrists, papules on bum/ genitalia. Treat all members of household at the same time.

Treat with aqueous malathion 0.5% liquid (24h) or permethrin 5% dermal cream (8-12h) and repeat 1/52 later.

Do not have a hot bath prior to application of treatment.

For infants, young children, elderly and immunocompromised apply also to areas above the neck – scalp, face and ears.

Ivermectin is available on a named patient basis for the tx of Norwegian scabies that is resistant to topical tx. Itching may persist for up to 6/52 after tx.

Tinea unilateral scaling and fissuring of one palm, feet, groin, and body.

Terbinafine is fungicidal (toenails) vs. clotrimazole is fungistatic. Tinea capitis (trichophyton) requires oral griseofulvin, terbinafine or itraconazole syrup 1/12.

Tinea versicolor requires topical imidazoles.

ACTINIC KERATOSIS (Celtic skin)
- Sunburn to forehead
- Ablative tx with solaraze (topical diclofenac), efudix BD x 6/52 or cryo (reoccurs as DNA remembers)
- Photodynamic therapy (PDT)

BULLOUS PEMPHIGOID (elderly)
- Autoimmune watery blisters
- Diagnosis made on biopsy
- Treat with prednisolone and azathioprine

- vs. pemphigus which is very rare

BOWEN'S DISEASE (SCC in situ)
- Solitary, painless, flat, occasionally itchy plaque on the lateral lower leg
- Treat with excision and graft or topical efudix.

DERMATOSIS PAPULOSA NIGRA (DPN) Small seborrhoeic warts found on the cheeks and neckline of Africans. Refer to dermatology for removal with cautery device.

ECZEMA (TYPES)
- **Atopic** – flexural and face. Treat with emollients, topical steroids +/- antibiotics for 2° infection. Advise parents to avoid man-made fibres in children's clothing and to buy natural fibres (cotton, linen, silk and wool. This also applies to bedding. Avoid scented bath additives, soaps, detergents or fabric softeners. New treatment –2ⁿᵈ line tacrolimus ointment (protopic) and does not cause skin atrophy.
- **Contact** – localised or unilateral. Occupational? Short course of steroids. Refer for patch tests.
- **Discoid** – circular lesions, especially on lower limbs, itchy, may weep with crusting. Treat with potent topical steroid +/- antibiotics. Stress may be trigger.
- **Irritant Hand Eczema** – finger webs and palms. Water damage. Treat with regular emollients and short periods of topical steroids.
- **Seborrhoeic** – red, scaly plaques on hairline, scalp, eyebrows, sides of nose and anterior chest. Treat with topical antifungal or 1% hydrocortisone, nizoral or capasal shampoo.
- **Varicose** – treat underlying venous disease with surgery or compression bandages (after Dopplers have excluded arterial disease). Try emollients and mild topical steroids.

GRANULOMA ANNULARE - Flat, smooth purple ring on back of hands, elbows and feet. Spontaneous resolution or treat with PUVA to decrease inflammation and mask pigment

KELOIDS- Mepiform dressing is available on an FP 10 – rehydrates and allows the keloid to soften and flatten.
Leave on until it drops off. Takes 2-3/12 to see effect. 10 dressings = £30. Alternative: silicon dressing.

- Inject with intradermal triamcinolone or offer scar revision surgery.

NECROBIOSIS LIPOIDICA (diabetes)
- Symmetrical on both lateral sides of shins
- Treat with intradermal triamcinolone every 4-6/52 (3-4 sessions total)
- Does not correlate with glycaemic control

PSORIASIS Trigger factors – stress, smoking, alcohol, anti-malarials, beta-blockers, lithium
- Initial mx – emollients, vitamin D analogue cream or ointment (Dovonex to the body and Curatoderm to the face and body). May use Alphosyl HC to face.
- Topical steroids should not be used alone, alternate with vitamin D analogue.
- Scalp psoriasis: Alphosyl or T-gel shampoo, add topical steroid lotion or vitamin D analogue-scalp application. If still problematic, try Cocois into scalp, leave overnight, follow by vitamin D or steroid scalp lotion to descale.
- Flexoral psoriasis: trimovate cream/ ointment daily for up to 3/52
- Palms and soles: dermovate or diprosalic (potent topical steroids) for 6/52
- Psoriatic arthropathy: diclofenac and refer for systemic therapy
- Refer to dermatologist for: failure to improve after 3-6/12; narrow band UVB phototherapy for widespread psoriasis; systemic therapy
- **Systemic treatment**: Methotrexate once wkly for psoriatic arthropathy; avoid ASA, NSAIDs affects levels, check LFTs; Retinoid (vitamin A); Cyclosporine (affects kidney and immunosuppression)

SKIN CANCER (also recognise photo of keratoacanthoma)
- **BCC** – most common type, pearly papule which ulcerates and bleeds. Tx: surgery or radiation.
- **SCC** – exophytic, indurated, hyperkeratotic firm nodule on lips or ears. Metastasizes. The elderly, smokers and renal transplant recipients (on IS drugs) are at risk.
- **Melanoma** – multiple colours, irregularity and ↑ in size. Stage with sentinel biopsy. Treatment with high dose interferon, surgery, or vaccine trials. Worse prognosis with vertical growth phase vs. radical growth phase (little metastatic potential).

Topical Drugs in Pregnancy (FDA)

Category A (no risk): nystatin

Category B (minimal risk): not C/I by manufacturer for use during pregnancy): Amphotericin B, Azelaic Acid, cicloprox, clindamycin, erythromycin (also **category C** – cannot rule out significant risk), haloprogin, hydroquinone agent, masoprocol, meclocycline, metronidazole gel or cream, mupirocin, naftidine HCL, oxiconazole nitrate, permethrin, terbinafine. OTC: bacitracin, benzoyl peroxide (also category C).

C/I in pregnancy: topical retinoids; topical carbonic anhydrase inhibitors (acetazolamide, dorzolamide, brinzolamide) glaucoma eye drops; avoid β-blockers (timolol, levobunolol, betaxolol, carteolol) or use in the lowest possible dose in the 1st trimester and discontinue 2-3 days prior to delivery to avoid β-blockade in the infant. Avoid antibiotics: chloramphenicol, gentamicin, neomycin, rifampin, tetracycline, and tobramycin.

International Diabetic Federation/WHO Diabetes 2006

Diabetes Mellitus: Fasting plasma venous glucose ≥ 7.0 mmol/l (126 mg/dl) or 2-hour OGTT with 75g glucose, plasma venous glucose ≥ 11.1 mmol/l (200 mg/dl).

Impaired Glucose Tolerance: Fasting plasma venous glucose < 7 mmol/l (126 mg/dl) and 2-hour OGTT ≥ 7.8 mmol/l (140 mg/dl) and <11.1 mmol/l (200 mg/dl).

Impaired Fasting Glucose: Fasting plasma venous glucose 6.1–6.9 mmol/l (110-125 mg/dl) and (if measured) 2-h OGTT < 7.8 mmol/l (140 mg/dl), after a 75g glucose load (otherwise IGT).

DIABETES (NICE Dec 2008)

Lifestyle management/non-pharmacological management
Management of obesity and smoking cessation
Dietary advice Provide in a form sensitive to the patient's needs, culture and beliefs, being sensitive to their willingness to change and the effects on their quality of life. Emphasise advice on healthy balanced eating that is applicable to the general population when providing advice to patients. Encourage high-fibre, low-glycaemic-index sources of carbs in the diet, i.e. fruit, vegetables, wholegrains and pulses; include low-fat dairy products and oily fish; and control the intake of foods with saturated and trans fatty acids).

Integrate dietary advice with a personalised diabetes management plan, including other aspects of lifestyle modification, such as ↑ **physical activity and losing weight**. Target for patients who are overweight, an initial body weight loss of 5–10%. Individualise recommendations for carbohydrate and alcohol intake, and meal patterns. ↓ the risk of hypoglycaemia should be a particular aim for a pt using insulin or an insulin secretagogue.

Advise that limited substitution of sucrose-containing foods for other carbohydrate in the meal plan is allowable, but that care should be taken to avoid excess energy intake. Discourage the use of foods marketed specifically for diabetes.

When patients are admitted to hospital as inpatients or to any other institutions, implement a meal-planning system that provides consistency in the carbohydrate content of meals and snacks.

Management of depression

When setting a target glycated haemoglobin (HbA$_{1c}$):

- involve the patient in decisions about their individual HbA$_{1c}$ target level, which may be > 6.5% set for type 2 diabetes in general encourage the patient to maintain their individual target unless the resulting side effects (including hypoglycaemia) or their efforts to achieve this impair their quality of life; offer therapy (lifestyle and meds) to help achieve and maintain the HbA$_{1c}$ target level

- inform a patient with a higher HbA$_{1c}$ that any ↓ in HbA$_{1c}$ towards the agreed target is advantageous to future health

- avoid pursuing highly intensive mx to levels of **< 6.5%.**

Measure the individual's HbA$_{1c}$ levels at:

- 2–6-monthly intervals until the BG level is stable on unchanging treatment; use a measurement made at an interval of < 3 months as a indicator of direction of change, rather than as a new steady state.

- 6-monthly intervals once the BG level and BG-lowering treatment are stable.

- If HbA$_{1c}$ levels remain > target levels, but pre-meal self-monitoring levels remain well controlled (< 7.0 mmol/litre), consider self-monitoring to detect postprandial hyperglycaemia (> 8.5 mmol/litre) and manage to lower this level if detected. Investigate unexplained discrepancies between HbA$_{1c}$ and other glucose measurements. Seek advice from a team with specialist expertise in diabetes or clinical biochemistry.

Offer Self-monitoring of BG to a pt newly diagnosed with type 2 only as an integral part of his or her self-mx ed. Discuss its purpose and agree how it should be interpreted and acted upon.

Self-monitoring of plasma glucose should be available: to those on insulin, to those on oral glucose-lowering meds to provide info on hypoglycaemia, to assess changes in glucose control resulting from medications and lifestyle changes, to monitor changes during intercurrent illness, to ensure safety during activities, i.e. driving.

Assess at least annually and in a structured way: self-monitoring skills, the quality and appropriate frequency of testing, the use made of the results obtained, the impact on quality of life, the continued benefit, the equipment used. If self-monitoring is appropriate but BG monitoring is unacceptable to the patient, discuss urine glucose monitoring.

Oral glucose control therapies (1): metformin, insulin secretagogues and acarbose Metformin Start in a patient who is overweight or obese (tailoring body-wt-assoc risk according to ethnicity) and whose BG is inadequately controlled by lifestyle interventions (nutrition and exercise) alone.

- Consider metformin as an option for 1st-line glucose-lowering tx for a patient who is not overweight.
- Continue with metformin if BG control remains or becomes inadequate and another oral glucose-lowering rx (usually a sulfonylurea) is added. (Metformin treats insulin resistance so may continue when adding tx.)
- Step up metformin gradually over weeks to minimise risk of GI side effects. Consider a trial of extended-absorption metformin tablets where GI tolerability prevents continuation of metformin tx.

Review the dose of metformin if creatinine > 130 micromol/litre or the eGFR <45 ml/minute/1.73-m^2.

Stop the metformin if the serum creatinine > 150 micromol/litre or the eGFR < 30 ml/minute/1.73-m^2.

Prescribe metformin with caution for those at risk of a sudden deterioration in kidney function and those at risk of eGFR falling < 45 ml/minute/1.73-m^2.

Discuss the benefits of metformin with a pt with mild to mod liver dysfunction or cardiac impairment so that: due consideration can be given to the CV-protective effects of the drug an informed decision can be made on whether to continue or stop the metformin.

Insulin secretagogues

Consider a sulfonylurea as an option for 1st-line glucose-lowering tx if: the pt is not overweight, does not tolerate metformin (or it is C/I) **or** a rapid response to tx is required because of hyperglycaemic sxs (metformin takes 1-2 weeks).

Add a sulfonylurea as 2nd-line when BG control remains or becomes inadequate with metformin.

Continue with a sulfonylurea if BG control remains or becomes inadequate and another oral glucose-lowering med is added. Prescribe a sulfonylurea with a low acquisition cost (but not glibenclamide) when an insulin secretagogue is indicated. When drug concordance is a problem, offer a once-daily, long-acting sulfonylurea.

Educate a patient on an insulin secretagogue, particularly if renally impaired, about the risk of hypoglycaemia.

Rapid-acting insulin secretagogues Consider offering to a patient with an erratic lifestyle.

Acarbose- Consider acarbose for a patient unable to use other oral glucose-lowering meds.

Oral glucose control therapies (2): other oral agents and exenatide

Glitazones, gliptins and exenatide, will be updated in the NICE short CG 'Newer agents for blood glucose in type 2 diabetes'. **Thiazolidinediones (glitazones)** If BGs are not adequately controlled (to HbA$_{1c}$ < 7.5%), consider adding a thiazolidinedione to the combo of metformin and a sulfonylurea if human insulin is likely to be unacceptable or ineffective because of employment, social or recreational issues related to putative hypoglycaemia, injection anxieties, other personal issues, or obesity/metabolic syndrome.

Consider adding a thiazolidinedione as 2nd-line therapy to:

- metformin as an alternative to a sulfonylurea (if HbA$_{1c}$ ≥ 6.5%) where the patient's job or other issues make the risk of hypoglycaemia with sulfonylureas particularly significant
- sulfonylurea monotherapy when BG control remains or becomes inadequate (HbA$_{1c}$ ≥ 6.5%) if the pt does not tolerate metformin (or it is contraindicated).

Warn a patient prescribed a thiazolidinedione about the possibility of significant oedema. Do not commence or continue a thiazolidinedione in patients who have evidence of heart failure, or who are at higher risk of fracture.

When selecting a thiazolidinedione for initiation and continuation of tx, take into account up-to-date advice from the relevant regulatory bodies (the European Medicines Agency and the Medicines and Healthcare products

Regulatory Agency), cost and safety issues (note that only pioglitazone can be used with insulin).

Gliptins – GLP-1 enhancers No recommendations are made as these oral drugs are not covered in this CG.

(sitagliptin or vildagliptin) DPP4 (GLP1 breakdown) inhibitors. 3rd line alternative to glitazone. ↓HbA1c by 0.5-1%.

Exenatide – GLP-1 mimetic Exenatide is NOT recommended for routine use in type 2 diabetes.

Consider exenatide as an option only if all the following apply:

- a BMI > 35.0 kg/m^2 in those of European descent, with appropriate adjustment for other ethnicities, specific problems of a psychological, biochemical or physical nature arising from high body weight
- HbA$_{1c}$ ≥ 7.5% with conventional oral agents after a trial of metformin and sulfonylurea
- other high-cost med, i.e. a thiazolidinedione or insulin injection, would otherwise be started.

Continue exenatide only if a beneficial metabolic response (at least 1.0% point HbA1c ↓ in 6 months and a wt loss of at least 5% at 1 year) occurs and is maintained. (Major side-effect is nausea for 1 month. Incretin, gut hormone which stimulates insulin release and ↓ gastric emptying (fullness) leading to wt loss. 3rd line instead of insulin. BD injection. Refer)

Glucose control: insulin therapy oral agent combination with insulin

When starting basal insulin: continue with metformin and the sulfonylurea (and acarbose, if used), review the use of the sulfonylurea if hypoglycaemia occurs. Studies of the role of sulfonylureas when starting a pre-mixed insulin preparation. Both pre-mixed insulins and sulfonylureas are effective glucose-lowering agents throughout the day, but can cause hypoglycaemia. When starting insulin, continuing sulfonylureas prevents deterioration of glucose control during insulin dose titration and reduces the requirement for insulin. However, it is not clear that these advantages are not offset by an ↑ risk of hypoglycaemia.

When starting pre-mixed insulin therapy (or mealtime plus basal insulin regimens): continue with metformin, continue the sulfonylurea initially, but review and discontinue if hypoglycaemia occurs.

Consider combining pioglitazone with insulin for: a pt who has previously had a marked glucose-lowering response to thiazolidinedione; a pt on high-dose insulin tx whose BG is inadequately controlled.

Warn the pt to discontinue pioglitazone if clinically significant fluid retention develops.

Insulin therapy (> 50% NIDDM will require and weight gain is drawback, i.e. stop 'peeing' the sugar)

When other measures no longer achieve a HbA$_{1c}$ < 7.5% or other higher level agreed, discuss the benefits and risks of insulin tx. When starting insulin therapy, use a structured programme employing active insulin dose titration that encompasses: structured ed, continuing telephone support, frequent self-monitoring, dose titration to target, dietary understanding, mx of hypoglycaemia, mx of acute changes in BG control, support from an appropriately trained and experienced healthcare pro. (Education: how to use insulin pen, sick day rules (when to call for help and adjusting doses), travel, keeping insulin in the fridge).

Preferably begin with human NPH insulin, taken at bed-time or bd according to need. (Humulin I or Insulatard)
Consider, as an alternative, using a **long-acting insulin analogue (insulin glargine)** for a patient who falls into 1 of the following categories: Those who require assistance from a carer or healthcare professional to administer their insulin injections. Those whose lifestyle is significantly restricted by recurrent symptomatic hypoglycaemic episodes. Those who would otherwise need bd basal insulin injections in combo with oral glucose-lowering meds.

Consider twice-daily biphasic human insulin (pre-mix) regimens in particular where HbA$_{1c}$ is > 9.0%. A once-daily regimen may be an option when initiating this treatment.
Consider pre-mixed preps of insulin analogues rather than pre-mixed human insulin preps when: immediate injection before a meal is preferred, hypoglycaemia is a problem, or there are marked postprandial BG excursions.
Offer a trial of insulin glargine if a pt started with NPH insulin has significant nocturnal hypoglycaemia.
Monitor using a basal insulin regimen (NPH or a long-acting insulin analogue [insulin glargine]) for the need for mealtime insulin (or a pre-mixed insulin preparation). If BG control remains inadequate (not to agreed target levels without problematic hypoglycaemia), move to a more intensive, mealtime plus basal insulin regimen based on the option of human or analogue insulins.
Monitor using pre-mixed insulin once or twice daily for the need for a further preprandial injection or for an eventual change to a mealtime plus basal

insulin regimen, based on human or analogue insulins, if BG control remains inadequate. Insulin detemir is not covered by this guideline.

Insulin delivery devices
Offer ed to a pt who requires insulin about using an injection device (usually a pen injector and cartridge or a disposable pen) that they and/or their carer find easy to use. Appropriate local arrangements should be in place for the disposal of sharps. If a pt has a manual or visual disability and requires insulin, offer a device or adaptation that: takes into account his or her individual needs.

Measure BP at least annually in a pt without previously diagnosed hypertension or renal disease. Offer and reinforce preventive lifestyle advice. For a patient on antihypertensives at dx of diabetes, review control of BP and meds used, and make changes only where there is poor control or where current meds are not appropriate because of microvascular complications or metabolic probs.

Repeat BPs within:
- 1 month if BP is >150/90 mmHg
- 2 months if BP is > 140/80 mmHg; 2 months if BP is > 130/80 mmHg and there is kidney, eye or CV damage.

Offer lifestyle advice (diet and exercise). Offer lifestyle advice if BP is confirmed as being consistently > 140/80 mmHg (or >130/80 mmHg if there is kidney, eye or CV damage). Add meds if lifestyle advice does not ↓ BP to < 140/80 mmHg (< 130/80 mmHg if there is kidney, eye or CV damage).

Monitor BP 1–2-monthly, and intensify treatment if on meds until BP is consistently < 140/80 mmHg (< 130/80 mmHg if there is kidney, eye or cerebrovascular disease).

1st-line BP-lowering treatment should be a once-daily, generic ACE inhibitor. Exceptions to this are African-Caribbean descent or F for whom there is a possibility of becoming pregnant.

1st-line BP-lowering treatment for a person of Afro-Caribbean descent should be an ACE inhibitor + either a diuretic or a generic ca-channel blocker. A ca-channel blocker should be the 1st-line BP-lowering treatment for a F for whom, after an informed discussion, it is agreed there is a possibility of her becoming pregnant.

For continuing intolerance to an ACEI (other than renal deterioration or hyperkalaemia), substitute an ATIIR antagonist.

If the BP is not ↓ to the individually agreed target with 1st-line tx, add a ca-channel blocker or a diuretic (usually bendroflumethiazide, 2.5 mg daily). Add the other drug (that is, the ca-channel blocker or diuretic) if the target is not reached with dual tx. If the BP is not ↓ to the individually agreed target with triple tx, add an α-blocker, a β-blocker or a K-sparing diuretic (the last with caution if the pt is already taking an ACEI or an ATIIR antagonist).

Monitor the BP of a pt who has attained and consistently remained at his or her BP target q 4–6 months, and check for adverse effects of antihypertensive tx – including the risks from unnecessarily low BP.

The use of ACEIs and angiotensin II-receptor antagonists in combination in early diabetic nephropathy.

Both of these classes of renin–angiotensin system blockers are effective in ↓ the rate of progression of diabetic kidney damage. However, there are acute risks of side effects associated with both classes. As these risks are similar, it is not clear whether the expected combined benefit from ACEIs and ATIIR antagonists would outweigh the combined risks.

Cardiovascular risk estimation

Consider a patient to be at high premature CV risk for his or her age unless he or she: is not overweight, tailoring this with an assessment of body-weight-associated risk according to ethnic group, is normotensive (< 140/80 mmHg in the absence of antihypertensive tx), does not have microalbuminuria, does not smoke, does not have a high-risk lipid profile, has no history of CVD **and** has no fhx of CVD.

If the patient is considered not to be at high CV risk, estimate CV risk annually using the UK Prospective Diabetes Study (UKPDS) risk engine. Perform a full lipid profile (including HDL and TG) when assessing CV after dx and annually, and before starting lipid-modifying treatment.

Management of blood lipid levels Statins and ezetimibe

Review CV risk status annually by assessment of CV risk factors, including features of the metabolic syndrome and waist circumference, and change in personal or family CV hx.

For a patient who is ≥ 40 years old: initiate generic simvastatin (to 40 mg) or similar, unless the CV from non-hyperglycaemia-related factors is low. If the CV risk from non-hyperglycaemia-related factors is low, assess CV risk using the UKPDS risk engine and initiate simvastatin (to 40 mg), or a similar statin, if the CV risk > 20% over 10 yrs.

For a patient who is < 40, consider initiating generic simvastatin (to 40 mg), or a similar statin, where the CV risk factor profile appears particularly poor (multiple features of the metabolic syndrome, presence of conventional RFs, microalbuminuria, at-risk ethnic group, or strong FHx of premature CVD). Once started on cholesterol-lowering tx, assess the lipid profile (with other modifiable RFs and any new dx of CVD) 1–3 months after starting treatment, and annually thereafter. In those not on cholesterol-lowering treatment, reassess CV risk annually and consider initiating a statin. ↑ the dose of simvastatin, in anyone initiated on simvastatin in line with the above recommendations, to 80 mg daily unless TC level is < 4.0 mmol/litre or LDL level is < 2.0 mmol/litre. Consider intensifying cholesterol-lowering tx (with a more effective statin or ezetimibe in line with NICE) if there is existing or newly dx of CVD, or if there is an ↑ albumin excretion rate, to achieve a TC level < 4.0 mmol/litre (and HDL not > 1.4 mmol/litre) or an LDL< 2.0 mmol/litre. If there is a possibility of a F becoming pregnant, do not use statins unless the issues have been discussed.

Fibrates

If there is a history of ↑serum TGs, perform a full fasting lipid profile when assessing CV risk annually.

Assess possible 2° causes of high serum TGs, including poor BG control (others include hypothyroidism, renal impairment and liver inflammation, i.e. alcohol). If a 2° cause is identified, manage according to need. Rx a fibrate (fenofibrate as 1st-line), if TG levels remain > 4.5 mmol/litre despite attention to other causes. In some circumstances, this will be before a statin has been started because of acute need (that is, risk of pancreatitis) and because of the undesirability of initiating two drugs at the same time. If CV risk is high (as is usual in type 2), consider adding a fibrate to statin treatment if TG levels remain in the range 2.3–4.5 mmol/litre despite statin tx.

Do not use nicotinic acid preps and derivatives routinely with type 2. They may have a role in a few patients who are intolerant of other tx and have more extreme disorders of blood lipid metabolism, when managed by specialist.

Omega-3 fish oils Do not prescribe fish oil preps for the 1° prevention of CVD in type 2. This recommendation does not apply to patients with hypertriglyceridaemia receiving advice from a specialist in blood lipid mx.

Consider a trial of highly concentrated, licensed omega-3 fish oils for refractory hypertriglyceridaemia if failed lifestyle and fibrate treatment.

Anti-thrombotic therapy

Offer low-dose aspirin, 75 mg od, if ≥ 50 years, if BP is < 145/90 mmHg.
Offer low-dose aspirin, 75 mg od, if < 50 years old and has significant other CV risk factors (features of metabolic syndrome, strong early fhx of CVD, smoking, hypertension, extant CVD, micro-albuminuria).
Clopidogrel should be used instead of aspirin only in those with clear aspirin intolerance (except in the context of acute CV events and procedures).

Kidney damage

Ask all with or without detected nephropathy to bring in a 1st-pass morning urine specimen yearly. In the absence of proteinuria/UTI, send this for lab estimation of albumin:creatinine ratio. Make the measurement on a spot sample if a 1st-pass sample is not provided (and repeat on a 1st-pass specimen if abnormal) or make arrangement for a 1st-pass specimen

Measure serum creat and estimate the GFR (using the method-abbreviated modification of diet in renal disease [MDRD] four-variable equation) **annually at the time of ACR estimation.**
Repeat the test if an abnormal ACR is obtained (in the absence of proteinuria/UTI) at each of the next two clinic visits but within a max of 3–4 months. Take the result to be confirming microalbuminuria if a further specimen (out of 2 more) is also abnormal (> 2.5 mg/mmol for men, > 3.5 mg/mmol for women).
Suspect renal disease other than diabetic nephropathy and consider further ix or referral when the ACR is raised and **any** of the following apply: there is no significant or progressive retinopathy, BP is particularly high or resistant to treatment, the patient previously had a documented normal ACR and develops heavy proteinuria (ACR > 100 mg/mmol), significant haematuria is present, the GFR has worsened rapidly, the patient is systemically ill.
Discuss the significance of a finding of abnormal albumin excretion rate, and its trend over time.
Start ACEIs with the usual precautions and titrate to full dose in all with confirmed raised albumin excretion rate (> 2.5 mg/mmol for men, > 3.5 mg/mmol for women). Have an informed discussion before starting an ACEI in a F for whom there is a possibility of pregnancy, assessing the relative risks

and benefits of the ACEI. Substitute an ATIIR antagonist for an ACEI for a patient with an abnormal ACR if an ACEI is poorly tolerated.

For a patient with an abnormal ACR, maintain BP < 130/80 mmHg. Agree referral criteria for specialist renal care.

Eye damage

Arrange or perform eye screening at or around the time of dx and annually. Use mydriasis with tropicamide when photographing the retina, after obtaining consent. Discussions should include precautions for driving.

Use a quality-assured digital retinal photography programme.

Perform VA testing as a routine part of eye surveillance programmes. Repeat structured eye surveillance according to the findings by: routine review in 1 year, or earlier review, or refer.

Arrange emergency review by ophthalmology for: sudden loss of vision, rubeosis iridis, pre-retinal or vitreous haemorrhage, retinal detachment. Arrange rapid review for new vessel formation.

Refer to an ophthalmologist in accordance with the National Screening Committee criteria and timelines if any of these features is present: referable maculopathy: exudate or retinal thickening within one disc diameter of the centre of the fovea, circinate or group of exudates within the macula (the macula is defined here as a circle centred on the fovea, with a diameter the distance between the temporal border of the optic disc and the fovea) any microaneurysm or haemorrhage within one disc diameter of the centre of the fovea, only if associated with deterioration of best visual acuity to 6/12 or worse, referable pre-proliferative retinopathy (if cotton wool spots are present, look carefully for the following features, but cotton wool spots themselves do not define pre-proliferative retinopathy):any venous beading, any venous loop or reduplication, any intraretinal microvascular abnormalities, multiple deep, round or blot haemorrhages, any unexplained drop in visual acuity.

Nerve damage assess with monofilament test

Diabetic neuropathic pain mx Make a formal enquiry annually about the development of neuropathic sxs causing distress. Discuss the cause and prognosis of troublesome neuropathic sxs, if present. Agree appropriate therapeutic options and review understanding at each clinical contact. Offer psychological support for the consequences of chronic, painful diabetic neuropathy. There is a need for comparison studies on tricyclic drugs, duloxetine, gabapentin and pregabalin which are partially effective as they differ in cost and side-effects. Use a tricyclic drug to treat neuropathic discomfort (start with low doses, titrated as tolerated) if standard analgesic measures have not worked, timing the med to be taken before the time of day when the sxs are troublesome; advise that this is a trial of tx. Offer a trial of duloxetine, gabapentin or pregabalin if a trial of tricyclic does not provide effective pain relief. The choice of drug should be determined by drug prices. Trials of these therapies should be stopped if the max tolerated drug dose is ineffective. If side effects limit effective dose titration, try another.

Consider a trial of opiate analgesia if severe chronic pain persists despite trials of other measures. If there is inadequate relief of the pain assoc with diabetic neuropathic sxs, seek assistance of a chronic pain management service.
If drug mx of diabetic neuropathic pain has been successful, consider ↓ the dose and stopping tx. If neuropathic sxs cannot be controlled adequately, discuss the reasons for the problem, the likelihood of remission in the medium term, the role of improved BG control. Gastroparesis Consider the diagnosis of gastroparesis in a patient with erratic BG control or unexplained gastric bloating or vomiting, taking into consideration possible alternative dx.

Consider a trial of metoclopramide, domperidone or erythromycin for an adult with gastroparesis. If suspect gastroparesis, refer to a specialist if: ddx is in doubt, or persistent or severe vomiting occurs.

Erectile dysfunction Review the issue of ED with men annually. Assess and address contributory factors and txs. Offer a phosphodiesterase-5 inhibitor, in the absence of contraindications, if erectile dysfunction is a prob. Following discussion, refer to a service offering other medical, surgical, or psychological mx of erectile dysfunction if phosphodiesterase-5 inhibitors are unsuccessful.

Other aspects of autonomic neuropathy Consider the possibility of contributory sympathetic nervous system damage for a patient who loses the warning signs of hypoglycaemia. Consider the possibility of autonomic neuropathy affecting the gut in an adult with unexplained diarrhoea, particularly at night. When using tricyclic drugs and antihypertensive meds in patients with autonomic neuropathy, be aware of the ↑ likelihood of s/e's i.e. orthostatic hypotension. Investigate unexplained bladder-emptying problems for the possibility of autonomic neuropathy affecting the bladder. Include in the mx of autonomic neuropathy sxs the specific interventions indicated by the manifestations (i.e., for abnormal sweating or nocturnal diarrhoea). **Pearls of wisdom** In a new symptomatic diabetic with nausea, vomiting, producing ketenes and with abdo pain, send to A+E! Act even if HbA1C is 7.6 as it will rise to 8-9 within 6 months!

Dx: 1 test if sxs, 2 tests if asymptomatic. FBG > 7.1 mmol/l; random BG > 11.1 mmol/l; IFG 6.1-7 mmol/l; IGT 7.8-11 mmol/l = arrange OGTT for IFG or IGT and if normal repeat annually.

Target as of 6/09, HbA1c 48-59 mmol/mol. Units changed from % to mmol/mol. 6% = 42, 7% = 53. 8% = 64, 9% = 75.

DIABETIC ORAL HYPOGLYCAEMIC DRUGS

CLASS DRUG	MODE OF ACTION	SIDE EFFECTS
α-glucosidase inhibitor		
Acarbose	delays absorption of CHO	flatulence, diarrhoea, jaundice
Biguanide		
Metformin	↓hepatic glc production (stop or can get lactic acidosis),	vomiting or diarrhoea **metallic taste**, abdo pain lactic acidosis (stop if eGFR < 30), ↓vitamin B12 abs

1st line BMI>25, max dose 1g bd, MR better tolerated
500 mg mane x 2/52, then 500 mg bd x 2/52, then 500 mg tds to avoid GI s/e's.

Insulin sensitisers		
Pioglitazone	↓insulin resistance and	GI upset, visual disturb, anaemia
(thiazolidinediones)	resensitise body to its own insulin	
Rosiglitazone		GI upset, anaemia. In US, ↑MI's with rosiglit.

Thiazolidinediones: cause fluid retention and wt gain so C/I in heart failure; ↑ jaw fractures.

Post-prandial glucose regulators		
Nateglinide	stimulate insulin secretion	hypersensitivity,
Repaglinide	by beta-cells	hypoglycaemia

Sulphonyureas (Give meglitinides for shift work.)

Gliclazide bd	stimulate insulin secretion	GI disturbance, hypersensitivity
If BMI< 23,	(short-acting) by beta cells	liver function dysfunction

Refer for early insulin.

Glibenclamide		**Avoid in elderly** (get confused, falls,
Glimepiride		especially with CKD)
Gliquidone		
Glipizide		
Tolbutamide (short-acting)		Headache, tinnitus
Chlorpropamide		**Avoid in elderly** (↑ risk of hypo's), facial flushing, ↑ADH secretion, most S/E's ↓Na

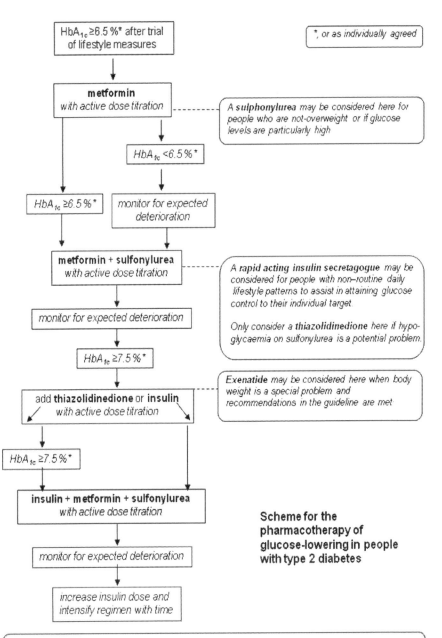

HbA$_{1c}$ ≥6.5 %* after trial of lifestyle measures

*, or as individually agreed

metformin
with active dose titration

A **sulphonylurea** *may be considered here for people who are not-overweight or if glucose levels are particularly high*

HbA$_{1c}$ <6.5 %*

HbA$_{1c}$ ≥6.5 %*

monitor for expected deterioration

metformin + sulfonylurea
with active dose titration

A **rapid acting insulin secretagogue** *may be considered for people with non–routine daily lifestyle patterns to assist in attaining glucose control to their individual target*

Only consider a **thiazolidinedione** *here if hypo-glycaemia on sulfonylurea is a potential problem.*

monitor for expected deterioration

HbA$_{1c}$ ≥7.5 %*

Exenatide *may be considered here when body weight is a special problem and recommendations in the guideline are met*

add **thiazolidinedione** or **insulin**
with active dose titration

HbA$_{1c}$ ≥7.5 %*

insulin + metformin + sulfonylurea
with active dose titration

Scheme for the pharmacotherapy of glucose-lowering in people with type 2 diabetes

monitor for expected deterioration

increase insulin dose and intensify regimen with time

For details see recommendations on
glucose-lowering targets, clinical monitoring, use of oral agents, and use of insulin

Measure BP annually if have no hypertensive or renal disease If >140/80 mmHg, confirm consistently raised

Targets: with retinopathy or cerebrovascular disease or with microalbuminuria
Follow algorithm with target <130/80 mmHg

Others
Follow algorithm with target <140/80 mmHg

↓ above target

Trial lifestyle measures alone unless >150/90 mmHg

Women with possibility of pregnancy avoid use of ACEI or A2RB drugs. Begin with CCB. In people with continuing intolerance to ACEI (other than renal deterioration or hyperkalaemia) substitute the ACE inhibitor with an A2RB drug.

↓ above target

| Maintain lifestyle measures | Start ACEI and titrate dose (if Afro-Caribbean add diuretic or CCB) |

People with microalbuminuria
Will already be on full dose ACEI or alternative
Then follow algorithm with target <130/80 mmHg

↓ above target

Add CCB or bendroflumethiazide

↓ above target

Add bendroflumethiazide or CCB

Scheme for the management of BP in type 2 diabetes
NICE 2008
ACEI, ACE inhibitor; A2RB, angiotensin 2 receptor blocker (sartan); CCB, calcium channel blocker

↓ above target

Add α-blocker, β-blocker, or potassium-sparing diuretic

↓ above target

Add α-blocker, β-blocker, or potassium-sparing diuretic, or refer to specialist

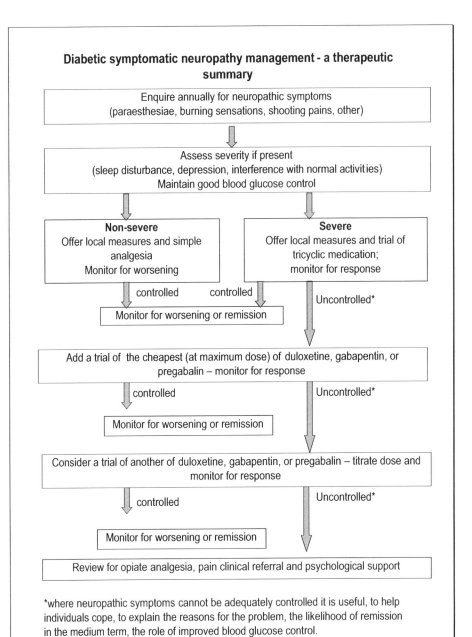

Diabetic symptomatic neuropathy management - a therapeutic summary

Enquire annually for neuropathic symptoms
(paraesthesiae, burning sensations, shooting pains, other)

Assess severity if present
(sleep disturbance, depression, interference with normal activities)
Maintain good blood glucose control

Non-severe
Offer local measures and simple analgesia
Monitor for worsening

Severe
Offer local measures and trial of tricyclic medication;
monitor for response

controlled   controlled   Uncontrolled*

Monitor for worsening or remission

Add a trial of the cheapest (at maximum dose) of duloxetine, gabapentin, or pregabalin – monitor for response

controlled   Uncontrolled*

Monitor for worsening or remission

Consider a trial of another of duloxetine, gabapentin, or pregabalin – titrate dose and monitor for response

controlled   Uncontrolled*

Monitor for worsening or remission

Review for opiate analgesia, pain clinical referral and psychological support

*where neuropathic symptoms cannot be adequately controlled it is useful, to help individuals cope, to explain the reasons for the problem, the likelihood of remission in the medium term, the role of improved blood glucose control.

DIETETICS: GLYCAEMIC INDEX
Know foods high in K+

LOW GI (< 55) - dairy (low fat/ fat free -milk, yoghurt, custard, cream); cereals (wholewheat, high fibre bran, oatbran, raw muesli, sultanas); bread and flour (provita crackers, seedloaf bread, pumpernickel bread); starches (tinned beans, peas, lentils, chickpeas, barley, rye, all pasta made from Durum Wheat, sweet potato, whole corn, basmati and tasmati rice); fruit (deciduous - apricots, cherries, peaches, pears, apples; citrus - oranges, grapefruit, lime, lemon; kiwi, grapes); most vegs; popcorn; sugar-free jam.

INTERM GI (55-70) - cereals (oatmeal, seminola, corn pops, shredded wheat); bread and flour (refined rye, oatmeal, ryevita); starches (sweetcorn, brown rice, baby/ new potatoes, couscous); fruit tropical - banana, mango, melons, pineapple; raisins, dates; veg (beetrooot, spinach); snacks/ sugars (pizza, homewheat digestive biscuits, low-fat (biscuits, muffins, pancakes), raw honey, jam, coke, fanta.

HIGH GI (> 70) cereals (muesli, pronutro, rice crispies, corn flakes, special K), starches (potato, all flour pasta, rice pasta); bread and flour (all brown, white and regular wholewheat bread, hot cross buns, melba toast; all flour - cornflour, breadflour, cakeflour, white scones and muffins, rice cakes); fruit (watermelon, dried fruit); veg (carrots, pasnips, pumpkin, turnips); snacks (sweets, wafer biscuits, commercial honey and syrup, tofu, energy drinks).

MEDIUM OXALATE

Apples/ Asparagus/ Artichokes/ Broccoli/ Brussel sprouts/ Butter/Carrots/ Corn/ Cucumber (med)/ Garlic/ Green beans, snap, or runner beans (?high)/ Kohlrabi/ Lettuce (Cabbage = low) iceberg/ Mushrooms/ Mustard greens/ Onions/Peppers green (1/2 medium) Potato chips (50) Potatoes, white, russet, Idaho (1/3 cup) (?high) Potato salad (1/4 cup)/ Radishes/ Snow peas/Tomato, fresh/ Tomato sauce, canned (1/4 cup)/ Veg beef soup /(Campbell's)/ Watercress.

HIGH IN OXALATE (> 10 mg per 1/2 cup serving)
Beans/ Beer/ Draft Beer/ Beets/ Beet greens/ Blackberries/ Gooseberries/ Carob powder/ Celery/ Chocolate/cocoa other chocolate drink mixes / Dark leafy greens/ Endive/ Escarole/ Eggplant/ Fruitcake/ Grits (white corn)/

Instant coffee (> 8oz/d)/ Legume types (baked beans canned in tomato sauce)/Leeks/ Nuts, nut butter (peanuts and pecans)/ Okra/ Parsley/ Peel: lemon, lime, orange/ Rasberries (black)/ Red currants/ Rhubarb/ Spinach/ Strawberries/ Summer squash/ Soy products (tofu, soy sauce)/ Sweet potatoes/ Swiss chard
Tea (including herbal) / Wheat bran/ Wheat germ/ Worcestershire sauce

SALT CONTENT IN COMMON FOODS
BJGP Vol 59 Issue 567: p786 K Mahtani

	Na (g)/serving (max 2.5g/day)	Salt (g)/serving (max 6g/day)
'Fast food' large hamburger	2.25	5.40
1 'high street' skinny blueberry muffin	0.60	1.54
Canned tomato soup (1/2 can)	0.60	1.50
'Fast food' medium fries	0.55	1.30
1 pizza slice	0.55	1.30
1 'high street' regular blueberry muffin	0.45	1.00
1 shake of salt cellar with multiple holes	0.40	1.00
1 rasher streaky bacon	0.36	0.90
1 processed slice of cheese	0.28	0.70
Wheat biscuit cereal (without milk)	0.24	0.60
Small packet crisps	0.20	0.50
1 slice brown bread	0.18	0.60
Peanut butter	0.15	0.40
Margarine	0.06	0.10
1 medium apple	< 0.01	< 0.01

DRUG MISUSE AND DEPENDENCE - Guidelines on Clinical Mx DOH Orange Book 1999

Prevalence: M: F ratio of 3:1. 55% heroin as main drug of misuse. Methadone, cannabis, amphetamine declining order of misuse. Injecting drug misusers are 22x more likely to die than non-injectors.

Treatment: ↓ risk of HIV, hepatitis B and C and other blood-borne infections from injecting.

- ↓ need for criminal activity. Reduce the use of illicit drugs by the individual.
- Assist the patient to remain healthy and stabilize patient on substitute medication.
- Shared care with GPs, specialist GPs, community psychiatric nurses, clinical psychologists, pharmacist, social worker and drug and alcohol workers.
- Specialist services for patients with dual dxs (mental illness + drug/ EtOH), liver disease, chaotic lifestyle, serious forensic hx, unresponsive to oral substitute rxs, specialized residential rehab programmes.

Assessment: Confirm patient is taking drugs (history, exam and urine toxicology screen).

- Assess degree of dependence (smoking, injecting, amount per week)
- Identify complications of drug misuse and assess risk behaviour (skin abscesses, h/o DVT)
- Identify medical, social and mental health problems.
- Offer advice on harm minimisation (clean needle exchange centres, HIV testing and hepatitis B immunizations).
- Determine the patient's expectations and degree of motivation.
- Assess the appropriate level of expertise needed for this patient (shared care).
- Determine the need for substitute medication such as methadone or subutex.
- Notify the patient to the local Regional Drug Misuse Database (form completion).

Hx and Physical Exam: Presentation: pregnant, impending court case, wanting help, etc.

- Past and current drug use: age at starting, types, quantity, frequency and routes of administration, overdoses, abstinences, symptoms (hallucinations, fits).
- History of injecting, risk of HIV and hepatitis.
- Med hx – abscess, DVT, chest infection, dental disease, TB, bacterial endocarditis, PE, LMP, smear, accidents.
- Psychiatric hx – admissions, OPC, overdoses, depression or concomitant psychosis (delusions or hallucinations), self-harm, attempted suicide.
- Forensic hx – probation, criminal record, outstanding charges
- Social hx – family, children., employment, accommodation (hostel), debt
- Past contact with treatment services – prior attempts at rehab, methadone, etc.
- Others – drug misuse in partner or family.
- Examine sites of injection for infection (neck, arms, groin and legs).

Ixs: Hb, creat, LFTs, hepatitis B and C, HIV Ab. Urine toxicology screen (opiates persist for 24 h, methadone 48h).

Prescribing doctor: No greater than one week's supply dispensed at one time. Keep clear written or computer records of rx.

- Monitor patient with regular urine toxicology tests to ensure compliance with substitute drugs.
- A multidisciplinary approach to drug treatment is essential.
- Prescribe methadone in opiate dependence and diazepam in benzodiazepine dependence. Lofexidine is useful for the treatment of withdrawal in a supervised community, inpatient, or residential setting.

Opiate withdrawal: Peaks at 36-72h. Presentation - sweat, lacrymation, rhinorrhoea, yawn, hot and cold, anorexia, abdominal cramps, tremor, N/V/D, insomnia, aches/ pains, dilated pupils, hypertension and ↑HR.

Detoxification for heroin addiction with methadone mixture 1 mg/ml (initial daily dose of 10-40 mg) or subutex (buprenorphine). 8 mg subutex is equivalent to 30 mg of methadone. Commence at 2 mg od.

DYSPEPSIA (NICE August 2004)

Review meds: calcium antagonists, nitrates, theophyllines, bisphosphonates, corticosteroids, NSAIDs.

Urgent specialist referral for endoscopy for patients of any age with dyspepsia and:

- Chronic GI bleed   Fe deficiency anaemia
- Progressive unintentional wt loss   Epigastric mass
- Progressive difficulty swallowing   Suspicious barium meal
- Persistent vomiting

Routine referral for endoscopy: if > 55 + sxs persist despite H pylori test and acid suppressing tx and when patients have ≥ 1 of: prior gastric ulcer or op, continuing need for NSAIDs or ↑ risk of gastric CA or anxiety re CA

- **Start with PPI x 1 month OR test for and treat H pylori** (carbon-13 urea breath test, stool antigen or lab-based serology). 2/52 washout period following PPI use before test for H pylori with a breath test or a stool antigen test. Rx 7-day, bd full-dose PPI + either metronidazole 400 mg + clarithromycin 250 mg OR amoxicillin 1g + clarithro 500 mg
- **GORD** - full-dose PPI for 1-2 mos. If symptoms recur, offer a PPI at the lowest dose and limit repeat rxs
- **PUD** - H pylori eradication if +. Stop NSAIDs. Offer full-dose PPI or H2RA rx for 2 mos.
- **Non-ulcer dyspepsia** - initial tx for H pylori if +, symptomatic mx and periodic monitoring

DYSPEPSIA Guidelines from the British Society of Gastroenterology 2002

Age for endoscopy 55yo. **Test and Treat:** Treat patients < 55 yo with uncomplicated dyspepsia on basis of a + H pylori
and not 'test and scope.' **13C Urea Breath Test** – the best test for ID and for confirmation of eradication of H Pylori. **Use of PPIs** – continue to follow NICE. **Common Causes**: Duodenal ulcer (+H Pylori) 10-15%; Gastric ulcer (+H Pylori) 5-10%; Oesophago/Gastric CA (+ H Pylori) 2%; Oesophagitis 10-17%; Gastritis, duodenitis (+H Pylori) or HH 30%; Normal 30%

Testing for HP

Serology has a high sensitivity but is less accurate than the urea breath test. Routine endoscopy for dx of H pylori is not recommended. **Endoscopy** for new onset uncomplicated dyspepsia, if > 55 (stop antisecretory therapy 4/52 before scope) + alarm sxs if < 55. **Endoscopy is inappropriate for**: DU which has responded symptomatically to treatment; patients < 55; patients who have recently undergone satisfactory endoscopy for same symptoms.

Alarm sxs: dysphagia and odynophagia, prior gastric op, epigastric mass, prior gastric ulcer, GI bleed, suspicious barium meal, persistent continuous vomiting, unexplained Fe deficiency anaemia, unintentional weight loss ≥ 3kg.

Tx for HP

Duodenal ulcer +HP– 1 week triple tx with PPI (bd) or RBC (ranitidine bismuth citrate) +amoxycillin 500g -1g bd or metronidazole 400-500 mg bd + clarithromycin 500 mg bd; quadruple tx for 2nd line tx with PPI + bismuth 120 mg qds + metronidazole 400-500 mg tds + tetracycline 500 mg qds

Duodenal ulcer no HP – cimetidine 800 mg nocte; refer GI if not NSAID-ulcer.

Gastric ulcer + HP – Heliclear + antisecretory therapy for 2/12. Long-term PPI or misoprostol if on NSAIDs.

Gastric ulcer no HP- antisecretory tx for 2/12. Stop NSAIDs. Give PPI if on NSAID.

Oesophagitis – 4/52 of antacids, raft preparations (alginate), H$_2$RA, or prokinetic agents (cisapride).

ECGs

Acute Pericarditis - Diffuse ECG changes in all leads except lead V$_1$ and aVR, Scooping/ concave ST segment – convexity is downwards, ST elevation, Blunted T wave. Aetiology: RA, SLE, Dressler's, Tb

MI Ant MI – Q's in V$_1$-V$_4$. Inf MI - Q's in II, III and aVF. Lat MI - Q's in I, aVL, V$_5$, V$_6$ and V$_2$ ST elevation or depression, ST segment is flat, scooped or shows coving (convexity is upwards)

Time scale: 0 mins	normal - left-sided leads always have Q waves
mins	peak T waves (ischaemia), hyperacute T waves
mins-hours	ST elevation with coving (injury)
mins-hours	T wave inversion
hours-days	Q (MI, necrosis) in an abnormal lead which is ≥ 0.4
sec	wide and contributes ¼ too much to QRS
days-weeks	Q wave remains, ST segment returns to normal, T wave inversion reverses

Pulmonary Embolism: S$_1$ Q Q$_3$ pattern, inverted T in leads V$_1$-V$_4$, ST dep in II, transient RBBB, RAD.

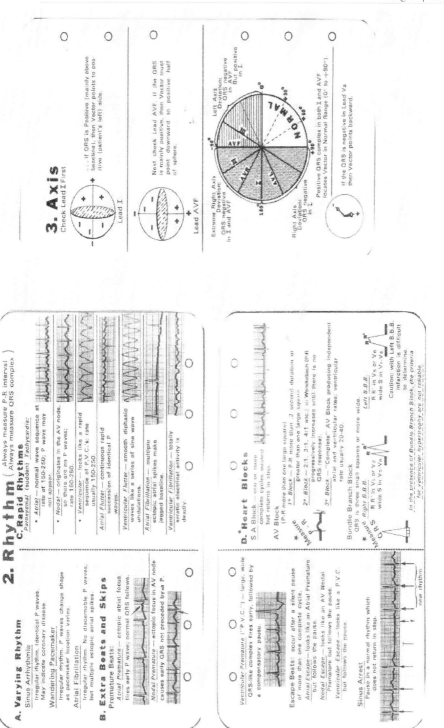

3. Axis
Check Lead I First

... if QRS is Positive (mainly above baseline), then Vector points to positive (patient's left) side.

Lead I

Next check Lead AVF. If the QRS is mainly positive, then Vector must point downward to positive half of sphere.

Lead AVF

Left Axis Deviation: QRS negative in AVF But positive in I.

NORMAL

Right Axis Deviation: QRS negative in I.

Extreme Right Axis Deviation: QRS negative in I and AVF.

Positive QRS complex in both I and AVF locates Vector in Normal Range (0° to +90°).

If the QRS is negative in Lead V₂ then vector points backward.

2. Rhythm (Always measure P-R interval / Always measure QRS complex)

A. Varying Rhythm

Sinus Arrhythmia
Irregular rhythm. Identical P waves. May indicate coronary disease.

Wandering Pacemaker
Irregular rhythm. P waves change shape as pacemaker location varies.

Atrial Fibrillation
Irregular rhythm. No discernable P waves, but multiple ectopic atrial spikes.

B. Extra Beats and Skips

Premature Beats:

Atrial Premature — ectopic atrial focus fires early P wave; normal QRS follows.

Nodal Premature — ectopic focus in AV node causes early QRS not preceded by a P.

C. Rapid Rhythms
Paroxysmal (sudden) *Tachycardia:*

* *Atrial* — normal wave sequence at rate of 150-250; P wave may not appear.

* *Nodal* — originates in the AV node, so there are no P waves; rate 150-250.

* *Ventricular* — looks like a rapid sequence of P.V.C.'s; rate usually 150-250.

Atrial Flutter — continuous rapid succession of identical P waves.

Ventricular Flutter — smooth diphasic waves like a series of sine wave undulations.

Atrial Fibrillation — multiple ectopic atrial spikes make jagged baseline.

Ventricular Fibrillation — totally erratic electrical activity is deadly.

Ventricular Premature ("P.V.C.") — large, wide QRS-like complex fires early, followed by a compensatory pause.

Escape Beats: occur after a silent pause of more than one complete cycle.

Atrial Escape — looks like an Atrial Premature but follows the pause.

Nodal Escape — looks like an AV Nodal Premature but follows the pause.

Ventricular Escape — looks like a P.V.C. but follows the pause.

Sinus Arrest
Pause in a normal rhythm which does not return in step.

D. Heart Blocks

S-A Block — one or more complete cycles missing but returns in step.

AV Block
(P-R more than one large square)

1° Block — P-R more than 1 second (duration of greater than 2 large squares).

2° Block — 2:1, 3:1, 4:1 etc.; or Wenckebach (P-R progressively increases until there is no QRS response).

3° Block — "Complete" AV Block producing independent atrial and ventricular rates. Ventricular rate usually 20-40.

Bundle Branch Block
QRS is three small squares or more wide.

Right B.B.B. — R,R' in V₁ or V₂; wide S in V₅ or V₆.

Left B.B.B. — R,R' in V₅ or V₆; wide S in V₁ or V₂.

Caution: with Left B.B.B. infarction is difficult to determine.

In the presence of Bundle Branch Block, the criteria for ventricular hypertrophy are not reliable.

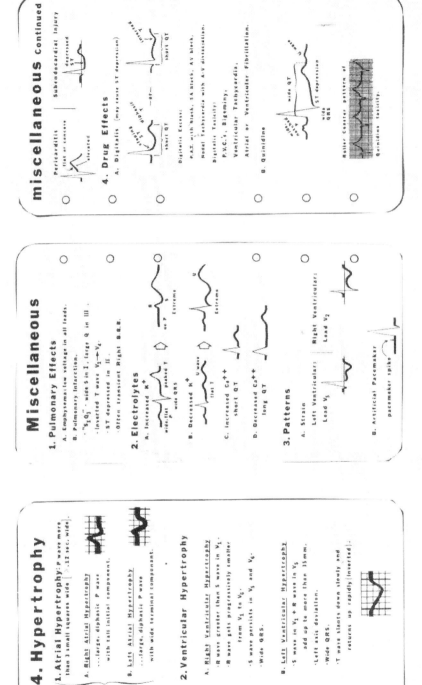

miscellaneous Continued

Pericarditis	Subendocardial Injury
flat or concave, elevated	depressed ST

4. Drug Effects (may cause ST depression)

A. Digitalis (may cause ST depression)

short QT   short QT

Digitalis Excess:
- P.A.T. with block, SA block, A-V block.
- Nodal Tachycardia with A-V dissociation.

Digitalis Toxicity:
- P.V.C.'s, Bigeminy,
- Ventricular Tachycardia,
- Atrial or Ventricular Fibrillation.

B. Quinidine

wide QT   ST depression   wide QRS

Roller Coaster pattern of Quinidine toxicity.

Miscellaneous

1. Pulmonary Effects

A. Emphysema: low voltage in all leads.

B. Pulmonary Infarction.
- "$S_1 Q_3$": wide S in I, large Q in III.
- Inverted T wave $V_1 \rightarrow V_4$.
- ST depressed in II.
- Often transient Right B.B.B.

2. Electrolytes

A. Increased K^+
- wide, flat P, peaked T, wide QRS

B. Decreased K^+
- U wave, flat T

C. Increased Ca^{++}
- short QT

D. Decreased Ca^{++}
- long QT

3. Patterns

A. Strain

Left Ventricular: Lead V_5

Right Ventricular: Lead V_2

B. Artificial Pacemaker: pacemaker spike

---

4. Hypertrophy

1. Atrial Hypertrophy: P wave more than 3 small squares wide (>.12 sec. wide).

A. Right Atrial Hypertrophy
...large, diphasic P wave with tall initial component.

B. Left Atrial Hypertrophy
...large, diphasic P wave with wide terminal component.

V_1

2. Ventricular Hypertrophy

A. Right Ventricular Hypertrophy
- R wave greater than S wave in V_1.
- R wave gets progressively smaller from V_1 to V_6.
- S wave persists in V_5 and V_6.
- Wide QRS.

B. Left Ventricular Hypertrophy
- S wave in V_1 + R wave in V_5 add up to more than 35 mm.
- Left axis deviation.
- Wide QRS.
- T wave slants down slowly and returns up rapidly (inverted).

ENT - Anatomy of the Right Tympanic Membrane

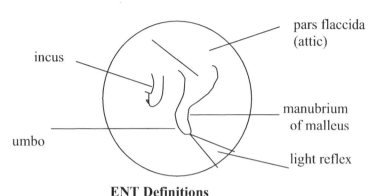

pars flaccida
(attic)

incus

manubrium
of malleus

umbo

light reflex

ENT Definitions

Acoustic neuroma – CN VIII tumour in the internal acoustic meatus. Presents with unilateral SNHL or tinnitus.

Attic Cholesteatoma – overgrowth of squamous epithelial cells from the attic of TM (pars flaccida) into the middle ear via an attic or posterior marginal TM perforation. The potential complication is erosion of bone, involvement of the dura and intracranial spread. Urgent referral. Surgical procedure: Mastoidectomy - limited by waiting list times of up to a year!

Aural polyp – chronic discharging ear. Associated with longstanding TM perforation and chronic suppurative otitis media. Surgical procedure: aural polypectomy and mastoid operation.

Bell's palsy – dx of exclusion; involves forehead (LMN: VII palsy); urgent ENT referral for OTA hearing test; rx eye drops and eye pad nocte; tx tapering dose of prednisolone 80, 60, 40, 20, 10 mg od.

Erysipelas – haemolytic strep enter through skin fissures or active otitis externa. Tx IV benzylpenicillin.

Facial palsy – congenital (forceps); metabolic (hyperthyroidism, malignant OE, pregnancy); infection (lyme disease, OE, OM, chickenpox, mastoiditis); neoplasm (cholesteatoma, parotid or basal cell CA, VII tumour); toxic (diphtheria, tetanus, thalidomide); trauma (head injury, temporal bone fracture, MVA); iatrogenic (post mastoidectomy, inferior dental block, rabies vaccine); idiopathic and neurological (Bell's palsy, Guillain-Barre, multiple sclerosis)

Grommet – plastic tube placed through an anterior-inferior TM myringotomy for treatment of glue ear
(secretory otitis media). Dislodges spontaneously after 12-18 months.

Herpes zoster (Ramsey-Hunt) – facial nerve palsy, vesicles in EAM

Malignant otitis externa – deep otalgia with granulation tissue blocking the EAM; found in diabetics; may affect CN VII-XII and lead to osteomyelitis, brain abscess and death; offending organism is pseudomonas pyocaneus, ix: CT scan + granulation tissue biopsy; tx with IV antibiotics.

Meniere's disease – nausea, vertigo, low-freq SNHL; with each attack, hearing not recover, may be bilateral.

Noise-induced hearing loss – recruitment, 4 Hz

Osteomas – swimmer's osteoma, bony hard swellings in the meatus, hyperostosis. Complications include wax build-up and otitis externa. Surgical intervention: removal with microdrill.

Otitis media – treat with ibuprofen if < 24h with fever and red TM. If > 24h, with T>40C and bulging red TM, treat with amoxicillin. Conductive hearing loss. Beware of post nasal space neoplasm in adults with unilateral secretory otitis media. Adenoidal hypertrophy leads to Eustachian tube dysfunction and otitis media with effusion infections in children.

Otosclerosis – conductive hearing loss with normal TM; commonly presents during pregnancy. Paracusis willisi – hears better in a noisy environment. Blue sclerae. Stapedectomy.

Presbyacusis – gradual bilateral high-frequency sensorineural hearing loss with age. Know PTA.

Secretory otitis media with retraction – golden or brown eardrum; prominent malleus; watchful waiting; 40 dB CHL on PTA with flat tymps; op: grommets + EUA PNS +/- adenoidectomy.

Ramsay-Hunt syndrome – HZ virus involves the geniculate ganglion of the VIIth cranial nerve; associated with facial palsy and SNHL. Tx: acyclovir 800 mg 5 times a day, analgaesia, Pope wick and steroid otic drops.

Tympanosclerosis incidental finding of chalk-like/ cotton-wool appearance to TM. ?PH of OM. No tx.

TUNING FORK TEST
Weber – (TF in middle of forehead – lateralises to side of CHL/ or opposite side of SNHL.

Do Rinne test next – louder in front of (AC> BC both sides) = Rinne positive = confirms opposite SNHL or behind the ear on mastoid (BC > AC) = Rinne negative = confirms ipsilateral CHL.

ENT (NOSE)
Nasal polyps
- benign grey opalescent 'peeled grapes'
- Conservative mx includes **betnesol nasal drops** or flixonase ampoules for 6/52; prednisolone 5 mg od for 2/52 or steroid nasal drops (beconase, nasacort, nasonex). Skin prick allergy testing is performed by the outpatient ENT nurse. Beware of the triad – aspirin sensitivity, nasal polyps and asthma. Cystic fibrosis is also associated with nasal polyps. Antrochoanal polyp arises within the maxillary antrum and protrudes into the nasopharynx.
- Indications for nasal polypectomy or FESS (functional endoscopic sinus surgery) – failed medical tx. A CT scan of the sinuses is obtained as the polyps arise from the ethmoid sinuses and can be associated with distortion of the nasal bridge and gross opacity of the maxillary and ethmoid sinuses. In this case a snare polypectomy is not adequate and the patient will require FESS/ polypectomy. Warn the patient of risk of injury to the orbit or brain.

Rhinitis (allergic)
- Refer for skin prick allergy testing.
- Prescribe betnesol drops 2 bd x 2/52, then give beconase spray (2 squirts bd) for months. Other sprays include my favourite, Nasacort (odourless, thixotropic – sticks to nasal mucosa, once a day with low

steroid content and effective within 16 hours, licensed for patients aged 6 and above) and nasonex. Rhinolast spray (antihistamine nasal spray) for children aged 2 to 12. Sodium cromoglycate (rhynocrom) good in kids. Beware flixonase (highest steroid content for efficacy).

- Prescribe antihistamines after topical treatment has failed.

Rhinitis (nonallergic; nonoesinophilic vasomotor)

- Drugs, hormonal, hypothyroidism, menopause, pregnancy, occupational and environmental irritants.
- Sympathomimetic rx (pseudoephedrine), topical nasal ipratropium spray (rinatec) for senile watery rhinorrhoea.
- Surgical option – reduction of inferior turbinates with diathermy.

Snoring

- Assess BMI, advise weight reduction if BMI > 30, ↓ EtOH intake, avoid heavy meals at night.
- If due to a deviated nasal septum, offer septoplasty.
- If due to a redundant uvula, offer uvuloplasty. If due to a redundant uvula and low hanging posterior pharynx, offer UVPPP. Surgical risks include pain, dry throat, voice change, etc.
- Otherwise arrange a sleep study to assess for apnoeic episodes if the BMI is not high.
- Patients with sleep apnoea must be advised not to drive.
- Children with sleep apnoea, may benefit from EUA and adenoidectomy. Some advocate concurrent tonsillectomy for obstructive airway.

Acute and chronic sinusitis

- Manage with antibiotics (2-6/52) and nasal decongestants (otrivine). Periorbital cellulitis requires urgent admission, IV antibiotics, urgent ophthalmologist review with visual acuity and colour vision check, and CT scan to exclude optic nerve compression. The patient may need urgent orbital decompression/ FESS with diminished visual acuity and affected colour vision.

BMJ 2003; 326:673 (29 March) **Benign paroxysmal positional vertigo**
BPPV is characterised by short-lived episodes of vertigo in associated with
rapid changes in head position. The pathology usually lies in the posterior SCC
of the inner ear. It is now widely accepted that "canalolithiasis" causes this
condition. Free floating debris in the endolymph of the SCC is assumed to act
like a plunger, causing continuing stimulation of the auditory canal for several
seconds after movement of the head has ceased. The condition is idiopathic in
most. The commonest identifiable cause, in some 20%, is minor trauma to the
head. The condition can present at any age but peaks in the 6th and 7th decades.

Patients with BPPV due to involvement of the posterior SCC typically have
episodic vertigo in association with a rapid change in head position,
particularly any movement relative to gravity. The vertigo lasts from a few
seconds to 1 min. Typical manoeuvres provoking vertigo include sitting up or
lying down in bed and turning to reach for objects on high shelves. Attacks
tend to occur in clusters, and may recur after an apparent period of remission.
The Hallpike manoeuvre is used to confirm the dx of BPPV due to
involvement of the posterior canal. A positive test provokes vertigo and
nystagmus when a pt is rapidly moved from a sitting to lying position with the
head tipped below the horizontal plane, 45° to the side, and with the side of the
affected ear (and SCC) downwards. Accompanying nausea may be intense.
The rotatory nystagmus typically has a latency of a few seconds before onset
and fatigues after 30-40 seconds. 2 main diagnostic pitfalls exist. Firstly,
patients who develop significant sxs with testing but do not develop
nystagmus do not have BPPV. Secondly, patients who have vertigo due to
pathology in the CNS may develop nystagmus with the Hallpike manoeuvre,
but typically this has no latent period, does not fatigue with time or repeated
testing, and is rarely accompanied by nausea. The spontaneous remission rate
for BPPV is high. In 1 RCT patients were recruited within 2 wks of the onset
of symptoms. 77% in the control group were significantly better after 1 month.

Vestibular suppressant and antiemetic meds are generally ineffective in BPPV.
In recent years tx has been greatly enhanced by the introduction of physical tx
which disperses the canal debris. **The Epley manoeuvre** entails a sequence of
movements of head and trunk to rotate the posterior SCC in a plane that
displaces the plug of debris from the canal into the utricle of the inner ear,
where it is inactive. A recent Cochrane review confirms the efficacy of the
Epley manoeuvre for treating BPPV. Pooled data from 2 trials comprising
86 patients yield an odds ratio of 4.92 (95% CI 1.84 to 13.16) in favour of tx

with resolution of sxs as an outcome. The odds ratio for conversion of a positive to negative Hallpike test is slightly higher at 5.67 (2.21 to 14.56). The status of instructions given to patients after tx is controversial. Anecdotally, many are advised to minimise head turning (if necessary with a soft collar) and sleep with their head raised on pillows, with the affected ear uppermost, for 48 hrs. Although this advice is based on a sound theory, there is no clinical evidence to support it.

BPPV can recur after successful tx. All the published trials focus on short term resolution of sxs as an outcome. There is no evidence to show that the Epley manoeuvre reduces later recurrence of BPPV, which is seen in the natural hx of the disease. However, patients who have frequent recurrences can be taught to perform the exercises themselves at home. A tiny proportion of patients who have severe recalcitrant symptoms may be considered for surgical tx- either surgery to obliterate the posterior SCC or singular nerve section. BPPV is a well defined clinical syndrome with a clear diagnostic test, and a safe, simple tx is available that takes 5 minutes to perform.

Epley's Manouevre to Treat BPPV

A. With patient sitting upright, turn head 45° towards affected side.
B. Quickly lie patient down just beyond the horizontal, maintaining head position. Hold for 30 seconds.
C. Turn patient's head to 45° above horizontal in the other direction. Hold for 30 seconds.
D. Roll patient onto side they are facing, so the nose is now pointing 45° below horizontal. Hold for 30 seconds.
E. Sit patient upright again.

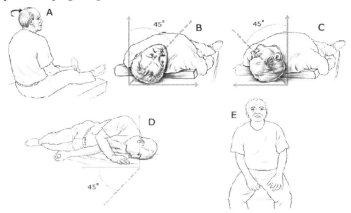

EPILEPSY
(NICE October 2004)

GPs to refer patients < 2 wks after 1st suspected seizure. Early referral as 15-30% of patients are misdiagnosed.

- Dx established by specialist. An EEG is only performed to support a diagnosis of epilepsy.
- MRI is the imaging ix of choice and is indicated for new onset epilepsy < age 2 or in adulthood, medication resistant seizure, and for patients with a focal onset suggestive on hx, exam or EEG.
- Appropriate ixs: serum lytes, gluc, calcium, 12-lead ECG and no longer include prolactin.
- Yearly review by GP. Information on SUDEP to patients.
- Treatment cessation is only advised if the patient has been seizure-free for 2 years.

Drug treatment of epilepsy

1st-line agents:
Sodium valproate (generalised tonic-clonic, absence, myoclonic, atonic and tonic, focal). Monitor for signs of blood, liver disorders or pancreatitis.

Carbamazepine (gen tonic-clonic, focal +/- 2^0 generalisation). Monitor for blood, liver or skin disorders.
Oxycarbazepine (focal +/-secondary generalisation). It is an analogue of carbamazepine with less induction of hepatic enzymes). Avoid abrupt withdrawal.
Lamotrigine (generalised tonic-clonic, absence, tonic, atonic, focal +/- secondary generalisation). Be wary for rash in children or signs of bone marrow failure. Monitor hepatic renal and clotting factors.
Topiramate (generalised tonic-clonic , myoclonic, focal +/- 2° generalisation, adjunctive therapy
for unresponsive seizure types)
Ethosuximide (1st-line for absence seizures in children)

2nd-line agents:
Vigabatrin (adjunct therapy for partial epilepsy).Monitor for visual field defects. Phenytoin and phenobarbitone (older broad-spectrum drugs with sedative and drug interactions).

Clonazepam (broad-spectrum second-line agent for absence, myoclonic, tonic and atonic).
Gabapentin (well-tolerated adjunctive therapy for focal +/-secondary gen)

Drug treatment of epilepsy in women on the pill

- Avoid hepatic enzyme inducers (carbamazepine, oxycarbazepine, phenytoin, phenobarbital and topiramate) in patients on coc, pop.
- Offer depo-provera with injection intervals of 10 weeks instead of 12 or offer non-hormonal IUD.
- If taking enzyme-inducing AED and elect to take the coc, then a minimum of 50 mcg of oestrogen is recommended. For BTB, increase oestrogen to 75-100 mcg or tricycle 3 packs (no breaks).
- POP and the progesterone implant are not recommended.
- If EC is prescribed, then 1.5 mg of levonorgestrel is required initially, then 750 mcg.

Drug treatment of epilepsy in pregnant women

- Avoid carbamazepine, phenytoin, sodium valproate - shown to be linked with neural tube defects.
- Prescribe 5 mg folic acid daily preconceptually and during the 1st trimester.
- Recommend use lamotrigine or gabapentin, which are not known to be teratogenic.
- Advise vitamin K1 prophylaxis against haemorrhagic disease of newborn if taking AEDs in last month of pregnancy.

BMJ Clin Evidence 2004 RCTs show immediate tx of a single seizure with AED ↓ recurrence at 2 yrs vs. no tx.

NSF guidelines for adult chronic conditions GPs should perform annual reviews of all well-controlled patients. This entails tailoring treatment options and discussing lifestyle issues such as ban on driving until fit-free for 12 mos, childcare issues, etc. Refer patient to epilepsy specialist nurse. Refer children for annual review by a specialist.

EQUAL OPPORTUNITIES LAW

The policy statement on equal opportunities and recruitment should be applied to all aspects of employment procedure to prevent discrimination in the workplace. It is important to consider the reasoning behind basic procedures and requirements and to take care with the wording of written procedures. In particular it is worthwhile considering how the recruitment process can be adapted to accommodate disabled applicants.

During recruitment and promotion, organisations should consider these points:

Job descriptions and person specifications for each vacant post should be drawn up or reviewed to **eliminate references to non-essential experience or qualifications** which might directly or indirectly discriminate against some candidates

Job advertisements should mention that the organisation is an equal opportunities employer and, where appropriate, through positive action encourage applications from ethnic minorities or other under-represented groups. Job advertisements should be displayed and promoted internally and externally, and be visible to all workers. They could also be placed in a **diverse range of press**, i.e., *Asian Times* and *The Voice*. The use of email and the internet can be an effective way of reaching different communities

Application forms sent to candidates should include a copy of the equal opportunities policy, job description and person specification. These papers could also be made available in large print and on computer disc.

The premises used for interview should be easily accessible for disabled candidates.

The timing of the interviews should be flexible to accommodate candidates with family commitments.

Interviewers should interview each candidate on the basis of the person specification.

Selection should be conducted solely on the basis of a candidate's relative merits and abilities.

The excellence of a worker in their present position should not be a reason not to promote.

The age, gender, sexual orientation, disability status, colour, race, religion, nationality, ethnic or national background of the candidates should be monitored by including a detachable questionnaire with the application form. This will enable the organisation, after selection, to determine the types of people who applied for a particular post and those who were shortlisted so that future advertising can be adjusted can be adjusted to avoid discrimination.

EYES
EXAMINATION:

- VA: near/ far, pinhole, +/- glasses
- colour vision: Ishihara (lost in optic neuritis, nerve transection or macular degeneration), red desaturation first (red colour is less bright)
- RAPD – relative afferent pupillary defect. Swing light across pupils. If the opposite pupil dilates paradoxically instead of consensual constriction to light, APD is present and indicates injury to the optic nerve. Present in optic neuritis.
- Do not dilate patients with angle closure glaucoma as can precipitate visual loss by sending the elderly pt with a small anterior chamber into angle closure.

AGE-RELATED MACULAR DEGENERATION
(degeneration of the macular leading to loss of central vision)

- **Dry type (80-95% of cases)** - non-exudative, painless, slow onset, loss of central vision over months or years. Drusen deposits and atrophy of retinal pigment epithelium. No available tx.
- **Wet type (10% of cases and 90% of sight loss in the UK) WET IS WORSE** - exudative, painless, rapid onset with progression of central vision loss. Straight lines appear to bend. +/- abnormal colour vision. Subretinal neovascularisaion with haemorrhage and scarring. PDT or Argon Laser only prevents further deterioration.

CATARACTS

- Lens opacity. See black shadow on lens at +10 diopter.
- Babies should be screened for cataracts. R/o metabolic disorder in children with cataracts.
- Implication of surgery for the pt is that the plastic lens cannot focus. So indication for surgery is if cataract prevents patient from doing what they want to do.
- Post cataract complications – acute ↓ in vision, with pain and red eye is endopthalmitis until proven otherwise. Poor red reflex. Refer promptly.

CHEMICAL INJURY:

- Apply LA and immediate wash out with saline/ water 1-2 L continuous irrigation for 15-30 mins.
- Check vision, F-stain, pH litmus paper (normal 7.5-8)

- Check cornea to exclude white eye which represent limbal ischaemia (blanched blood vessels) following alkali injury. This can progress to corneal necrosis within 24h. Refer urgently.
- If the entire eye stains blue and the eye is white, the patient will need grafts for weeks.
- Cotton bud sweep conjunctiva under lid to wipe off any metal or grit

DIABETIC RETINOPATHY

- Screen pregnant women with diabetes every 4-6 weeks to exclude maculopathy.
- Screen asymptomatic diabetics every 6/12 to 1 year.
- **Degree 1: background retinopathy (BDR)** – microaneurysms, no treatment, annual review
- **Degree 2: preproliferative retinopathy** – cotton wool spots marker for ischaemia, i.e. soft exudates at risk of developing new vessels; **maculopathy**
  - yellow hard exudates (lipoprotein) - treat with laser same day as seen in eye clinic
  - macula oedema – leakage of fluid, treat with laser to decrease visual loss by 50%
  - ischaemic (no blood supply) – no treatment
- **Degree 3: proliferative retinopathy** – new vessels can bleed and result in vitreous haemorrhage or become scar tissue causing traction on the retina and resultant visual loss. Panretinal photocoagulation is used to treat neovascularisation.

LIDS

- **Dacrocystitis** – tx broad-spectrum antibiotics for pus in lacrimal sac. May need dacrocystorhinostomy.
- **Ectropion** – routine referral for minor op.
- **Entropion** – invert the lid, if corneal transparency is gone, this is an ominous sign and the pt will need to be operated on within a week.
- **HZ ophthalmicus** – rarely crosses the midline. May see scabs around eye. If the nose is spared, then the eye is spared. Tx acyclovir, but if not, refer urgently as may develop CN III palsy + iritis.
- **Meibomian cyst** – the size of a marble may be removed under LA in OPC
- **Trichiasis** - inward turning of the eyelashes. Common in Oriental children (epiblepharon). Causes corneal scarring, abrasion and keratitis.

- **Thyroid disease** – overaction of Muller's muscle with uniocular proptosis.

PALSY

- In long-sightedness, a convergent squint occurs when not wearing glasses.
- **Lateral rectus palsy (CN VI)** – consider brain tumour, stroke. Limited lateral deviation of eye. May see convergent squint.
- **Sudden onset of painful diplopia, CN III palsy** – cannot look up and out or down and in, eye down and out with lid palsy and dilated pupil, consider compressive lesion (posterior communicating artery aneurysm or subarachnoid haemorrhage) and refer for urgent MRI. Diabetes does not affect the pupil.
- **Sudden onset of CN III palsy in a child** with ptosis and a lateral diverging eye, consider posterior communicating aneurysm.
- Superior oblique (CN VI) palsy – eye keeps going down and out on accommodation.
- Bilateral horizontal nystagmus – consider multiple sclerosis.
- Squint in children – Do cover test. Use torch and alternate. This will bring out the squint. Orthoptic referral to avoid amblyopia (useless vision).

RED EYE

Conjunctivitis

- Allergic (itchy and watering) – sodium cromoglycate (opticrom) eye drops
- bacterial – irritation with sticky discharge. Check for preauricular LNs. Tx chloramphenicol or fucithalmic drops.
- chlamydial (nonresolving conjunctivitis – consider testing)
- viral - most common and resolves in 1-2 wks; **adenovirus** (pharyngoconjunctival – fever with photophobia, self-limiting, resolves in 2-6/52, mandatory 2/52 off work as highly contagious)
- check vision and ideally do F-stain of cornea
- pupils equal and central, no photophobia, normal VA

Corneal abrasion/ dendritic HSV ulcer

- Contact lens wearer at risk – refer if has red eye.
- Do not apply local anaesthetic to a patient with a corneal abrasion, as the patient may rub his eye and make the abrasion larger.

- Recurrent abrasions common in young mothers from finger poking, postmenopausal F and in morning after alcohol-binge (dehydration). Treat with lubricants (hypromellose) for 2/12.
- Use F-dye to stain corneal break 'green' and use a blue light to pick up dendritic HSV ulcers 'green Xmas tree.' Do not use steroid eye drops, as this proliferates HSV and prevents graft uptake if needed.

Episcleritis – tender, localised to one segment of the eye. Treat with oral NSAIDs.

Scleritis – severe pain. Requires systemic steroids. Associated with arthritides.

Acute Iritis/ uveitis – acute red eye, painful (ciliary muscle) with photophobia and an irregular pupil as iris gets stuck. Worsening vision. Ask if the patient has had this before, as the differential includes HLA B27 arthritides, HIV, lyme disease, inflammatory bowel disease, idiopathic, sarcoidosis, toxoplasmosis. Tx: topical steroids and mydriatic drops (cyclopentolate) to rest pupil.

Acute angle closure glaucoma – usually elderly F presents at dusk when pupil dilates and closes off angle. Sxs include pain, headache, N/V, decreased VA, cloudy cornea, dilated and sluggish/ fixed pupil. The eyeball is rock hard on palpation, the pupil fixed, the conjunctiva injected and the vision diminished. Normal IOP is between 10 and 21. May see a sudden ↑ of IOP from 20 to 60, as aqueous humour cannot get in front of the iris and so gets occluded behind the iris. The iris then adheres to the back of the cornea. The optic nerve can be damaged within 6 hrs. Refer urgently to specialist. Administer antiemetic and IV acetazolamide (diamox) to ↓ aqueous humour production and ↓ IOP and release block with pilocarpine drops. Pt has 6-8h before she loses her vision.

FLOATERS

- If gradually increasing think vitreous degeneration, lens opacity, corneal scar.
- If sudden onset with flashing +/- ↓ in vision, think vitreous detachment, embolus in disc, blocked blood vessel or retinal detachment.
- 2/52 after cataract operation with counting fingers may represent infection, vitreous detachment or haemorrhage, or retinal detachment (dilated fundus exam to exclude). Check for RAPD.

GLAUCOMA

- Asymptomatic until visual fields are impaired.
- Associated with IOP > 21, optic disc cupping and visual field defects.
- For repeat prescriptions for timolol ask the patient if he is having SOB or eye soreness. If he is having beta-blocker side-effects, give xalatan nocte instead.

ORBITAL INJURY

- Check VA through a pinhole. ↓ VA may indicate presence of hyphaema or retinal oedema.
- Check optic nerve by checking VA and colour vision.
- If patient has diplopia on looking up and down, orbital blow out fracture? Palpate rim for notch of fracture.
- Fixed and dilated indicates ↑ IOP
- Small pupil suggests iritis.
- Check EOMI. Check for hyphaema and exclude retinal detachment or vitreous haemorrhage.
- Cover test
- Check for sensation with cotton ball under the eye (infraorbital nerve). Check for proptosis, enopthalmos
- Advise patient not to blow his nose as this may cause orbital inflation. Check IOP.

PINGECULA - yellow fatty deposit on eye; refer for cosmetic removal

PTERYGION - broad-based at 3 or 9 o'clock; takes 50-80 yrs for pterygia to cover iris. Routine referral for removal and conjunctival graft

SIDE-EFFECTS OF DRUGS

β-blocker   dry eyes
Prednisolone   glaucoma, papilloedema, posterior subcapsular cataract, corneal thinning
Rifampicin   orange-red tears

SUDDEN PAINLESS LOSS OF VISION

- Take a history and ask for sxs of giant cell arteritis, amaurosis fugax, curtain like effect with flashes or floaters (retinal detachment), diabetes +/- laser treatment. Check vision, visual fields, RAPD and fundus.

Anterior ischaemic optic neuropathy
- Elderly; headache; jaw claudication; temporal tenderness; polymyalgia
- swollen optic disc on fundoscopy;
- check ESR and CRP (more sensitive); risk of bilateral visual loss; treat with steroids

Central retinal artery occlusion
- History of atrial fibrillation, giant cell arteritis, hypertension
- cherry red spot in macula and swollen optic disc
- no light perception; APD
- Auscultate for carotid bruits.
- Refer immediately for treatment within 6-12 h to avoid optic nerve atrophy.
- Treat with IV diamox, ocular massage, AC paracentesis.

Central retinal vein occlusion
- finger counting; large optic disc; stormy sunset with engorged veins
- possible improvement to peripheral vision in 6-12 months
- at risk for new vessel formation; no definitive treatment

Optic neuritis (pain on movement)
- age 20-45 yo; ↓ vision and colour vision; RAPD; +/- disc swelling; unilateral
- seen in multiple sclerosis, no definitive treatment

Retinal detachment (urgent referral)
- 3 F's – flashers (flashing lights), floaters (look through frog spawn), field defect (1/2 of field)
- Get the pt to lie flat (if cannot see down below and see flashing lights, suggests upper detachment)
- RFs: high myopia, hypertension, sickle cell, prior ocular ops, diabetes, melanoma, breast CA, leukemia

Vitreous haemorrhage
- common in diabetics with neovascularisation; seen with retinal detachment and bleeding disorders
- spontaneous absorption; treatment: photocoagulation of new vessels

Anatomy of the visual paths and visual field correlation

Homonymous hemianopsia (Adult): 42-89% infarct and intracranial haemorrhage (stroke), #2 tumour, #3 trauma, surgical intervention; uncommon: MS, encephalitis, abscess, Creutzfeld-Jacob disease, seizures. (Child): CA.

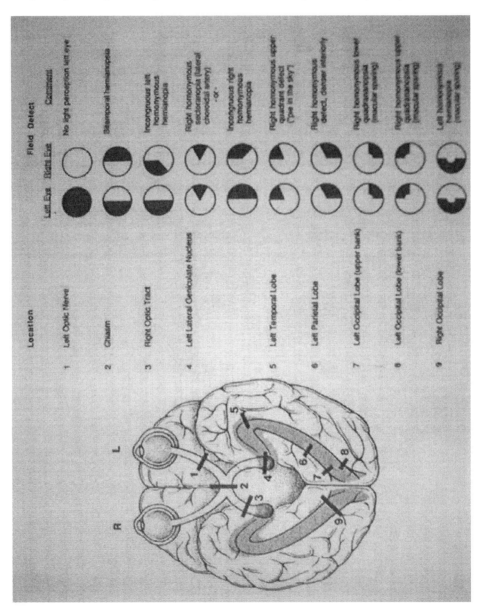

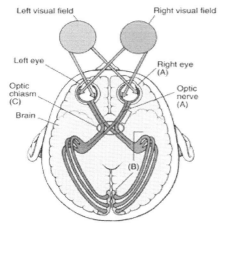

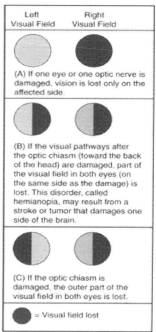

(A) If one eye or one optic nerve is damaged, vision is lost only on the affected side.

(B) If the visual pathways after the optic chiasm (toward the back of the head) are damaged, part of the visual field in both eyes (on the same side as the damage) is lost. This disorder, called hemianopia, may result from a stroke or tumor that damages one side of the brain.

(C) If the optic chiasm is damaged, the outer part of the visual field in both eyes is lost.

= Visual field lost

FERTILITY (NICE Feb 2004)

- 84% of couples conceive w/n 1 yr. Fertility ↓ with age. 94% aged 35, 77% aged 38 after 3 yrs.
- Screen for chlamydia before uterine instrumentation. F < 1-2 units 1x or 2x a week, M < 3-4 units/d.
- Offer folic acid supplemement before conception and up to 12/40. 5 mg od if on antiepileptic rx.
- Offer rubella susceptibility screening and vaccination.
- Definition Infertility: failure to conceive after regular UPSI for 2 years in the absence of reproductive pathology.
- If unable to conceive after a year, offer semen analysis +/or assess ovulation. Couples seen together.

Semen Analysis: volume ≥ 2 ml; liquefaction time: w/n 60 mins; pH ≥7.2; sperm concentration 20 million per ml; total sperm no: 40 million per ejaculate; 50% motility or more; vitality 75%; wbc < 1 million per ml; morphology: 15% or 30%. Screening for antisperm antibodies should NOT be offered as there is no effective tx to improve fertility. If the result of 1ˢᵗ semen is abnormal, repeat test 3months after initial analysis to allow time for cycle of spermatozoa formation to be completed. If azoospermia, repeat ASAP.

Assessing ovulation

- Offer serum progesterone in the mid-luteal phase (day 21 of a 28-day cycle) to confirm ovulation in women with regular menses and > 1 yr infertility. If prolonged menses, offer test later in cycle day 28 of 35-day cycle and repeat weekly until the next menstrual cycle starts.
- Use of basal body temp is NOT recommended. Routine TFTs is NOT recommended.
- Irregular menstrual cycles - offer serum FSH and LH tests.
- Measure prolactin ONLY if have an ovulatory disorder, galactorrhoea or a pituitary tumour.
- Tests of ovarian reserve have limited sensitivity and specificity. Using Inhibin B is uncertain in assessing reserve.
- Endometrial biopsy should NOT be offered to evaluate the luteal phase.

Assessing tubal damage

- Offer HSG to screen for tubal occlusion if not known to have co-morbidities (PID, previous ectopic or endometriosis)
- Hysterosalpingo-contrast-ultrasonography is an effective alternative.
- If known to have co-morbidities, offer laparoscopy and dye.

Assessing uterine abnormalities - NOT offer hysteroscopy on its own as part of initial Ix.
Postcoital testing of cervical mucus is NOT recommended.

Medical and surgical mx of male factor fertility problems

- Offer gonadotrophin drugs to men with hypogonadotrophic hypogonadism.
- Idiopathic semen abnormalities should NOT be offered anti-oestrogens, gonadotrophins, androgens, bromocriptine or kinin-enhancing drugs.
- Effectiveness of systemic steroids for antisperm antibodies is uncertain.
- Men with leukocytes in their semen should NOT be offered antibiotics without identified infection.
- Surgical correction of epididymal blockage for obstructive azoospermia.

- Surgery for varicocoeles should NOT be offered, as it does not improve pregnancy rates.

Ovulation Induction

- 21% of female fertility problems. WHO Group I: hypothalamic pit failure (hypothalamic amenorrhoea or hypogonadotrophic hypogonadism); Group II: hypothalamic pit dysfunction (PCO); III: ovarian failure.

Antioestrogens

- PCO should be offered clomifene citrate (or tamoxifen) as 1st line for up to 12 months to induce ovulation and told of risk of multiple pregnancies. Offer U/S monitoring during 1st cycle.

Metformin - is NOT currently licensed for the tx of ovulatory d/o.

- Anovulatory F with PCO who have not responded to clomifene and have a BMI > 25 should be offered metformin combined with clomifene. Warned of side-effects: nausea, vomiting, GI.

Laparoscopic ovarian drilling - PCO and not respond to clomifene.

Gonadotrophins

- PCO and not ovulate on clomifene, may offer human menopausal gonadotrophin, urinary FSH and recombinant FSH.
- If on clomifene and not pregnant after 6 mos, offer clomifene citrate-stimulating intra-uterine insemination.
- Used following pit-down regulation as part of IVF
- PCO and on gonadotrophins should NOT be offered GnRH agonist as associated with ↑risk ovarian hyperstimulation.
- GnRH agonists in long protocols are routinely used during IVF.

Use of GHs as an adjunct to ovulation induction tx is NOT recommended.

Pulsatile Gn releasing hormone should be offered to Group I ovulation d/os.

Dopamine agonists - bromocriptine for hyperprolactinaemia.

Tubal microsurgery and laparoscopic tubal surgery

For proximal tubal obstruction, offer selective salpingography+tubal cath or hysteroscopic tubal cannulation

Hysteroscopic adhesiolysis for intra-uterine adhesions.

Medical and surgical mx of Endometriosis

- Medical tx (ovarian suppression) should NOT be offered.
- For min or mild endometriosis, offer surgical ablation or resection of endometriosis + lap adhesiolysis.
- Ovarian endometriomas should be offered laparoscopic cystectomy.
- Moderate or severe endometriosis offer surgery. Post-op medical

treatment does not improve pregnancy rates.

Intra-uterine insemination

- Mild male factor, unexplained fertility, or min to mild endometriosis, offer up to 6 cycles.
- Ovarian stimulation should NOT be offered because of risk of multiple pregnancy.
- Single rather than double insemination should be offered.
- Fallopian sperm perfusion for insemination should be offered for unexplained fertility

Offer salpingectomy for women with hydrosalpinges.

IVF: Chances of live birth per IVF tx cycle: > 20% 23-35yoF; 15% 36-38yo; 10% 39 yo; 6% 40=>yo

- Screen for HIV, hepatitis B and hepatitis C.
- Adversely affect success rate of IVF - EtOH > 1 unit /day, smoking, caffeine, BMI > 30.
- Couples in which the F is 23-39 yo at time of tx and who have an identified cause of infertility(azoospermia or bilateral tubal occlusion) or who have infertility of minimum 3 years should be offered up to 3 stimulated IVF cycles.
- Human menopausal gonadotrophin, urinary FSH and recombinant FSH are equally effective in achieving a live birth when used following pituitary down-regulation as part of IVF.
- Inform couples of chance of multiple pregnancy after IVF depends on the number of embryos transferred per cycle of tx. No more than 2 embryos should be transferred during any one cycle of IVF.
- Embryos not transferred during a stimulated IVF treatment may be suitable for freezing. If ≥ 2 embryos are frozen then they should be transferred before the next stimulated tx cycle to minimise ovulation induction and egg collection, both of which carry risks for the patient + uses more resources.
- Clomifene citrate-stimulated and gonadotrophin-stim IVF have higher pregnancy rates than natural cycle

Intracytoplasmic sperm injection - severe deficits in quality, obstructive azoospermia, non-obstructive azoospermia; genetic counselling, karyotype testing. Testing for Y chromosome microdeletions NOT routine ix.

Donor insemination - obstructive azoospermia, non-obstructive azoospermia, HIV male, severe rhesus isoimmunisation, severe deficit in semen quality, risk of genetic disorder of offspring.

Oocyte donation - premature ovarian failure, Turners, bilateral oophorectomy, ovarian failure after chemo or radiotherapy, certain IVF fails
Application of cryopreservation in cancer treatments.
INFERTILE CRITERIA – refer to urology:
< 1 mill/ejaculate; motility < 20% (should be 40%); progression < 2/4 (twitch vs. dash), abnormal forms > 85%.
 IVF/ICSI NOT offered if F > 34/35 yo. 28.2% live births/cycle if F < 35 yo. ICSI is £5-7k/ cycle.
Oligozoospermia - < 20m/ml Mx: varicocoele ligation, antisperm Abs, IUI, or ICSI; Azoospermia = 0m/ml.

FITNESS TO DRIVE – DVLA GUIDELINES
updated August 2011 (www.dvla.gov.uk)

Condition	Ordinary car licence Group 1	LGV or PCV Group 2
Age	> 70 yo renew with completion of medical questionnaire every 3 yrs	Medical confirmation of fitness from ages 18- 45, then 5-yearly until 65, then annually.
AIDS	1, 2, 3 year licences with medical review	Maintain CDT count ≥ 200 for min 6 months. Assess case on individual basis.
HIV positive	Need not notify DVLA.	Need not notify DVLA.
Alcohol	Misuse – 6/12 revocation Dependency – 1year Hepatic cirrhosis – revoke Breath > 87.5 mcg/100 ml or BAL > 200 mcg/100ml	Off 1 yr for misuse Dependency- 3 yrs EtoH related d/o – revoke
Drugs	Cannabis, LSD – 6/12 Heroin, Cocaine -1 yr off drugs	1 year drug free 3 years drug free
Arrhythmia	Cease unless cause controlled for 4/52.	Disqualifies unless controlled for 3/12 with LVEF ≥ 0.4
Pacemaker/ Angioplasty	1/52 cease	Disqualifies for 6/52.
Implantable Cardioverter Defibrillator	6/12 after first implant	Permanently bars

Angina at rest/ Wheel	Cease until control. Need not notify.	Revocation. Relicense when angina-free for 6/52.
CABG	Cease for 4/52. Need not notify.	Disqualifies for 3/52. Must meet exercise test requirements & LVEF $\geq 40\%$.
ACS (MI)	Need not notify. After non STEMI, then 1/52 after angioplasty. Or 4/52 if no angio tx.	Disqualifies for 6/52. Relicense after exercise Test.
Aortic Aneurysm	Notify if > 6cms. Stop if $\geq$ 6.5 cms.	Disqualifies if > 5.5 cms
Asthma, COPD	Need not notify unless associated with LOC	Need not notify unless assoc. with LOC.
Obstructive sleep apnoea	Cease until controlled.	Cease until controlled.
Cancer	Need not notify unless brain mets. (If malignant brain CA, stop for 1 yr after Grade 1 & II glioma, else 2 yrs)	Assess cases on individual basis. Cease for 2 yrs after CA lung, no brain mets.
Craniotomy (meningioma)	off 6/12 postop	Refusal and revocation Relicense 5 yrs fit free
CVA/TIA	off 1 month (notify if residual deficit)	Revocation 12 mos after CVA or TIA
Diabetes	Notify DVLA if on tablets or insulin. Advice – warning signs of hypoglycaemia 1,2,3 yr licence.	From 15/11/2011, may no hypos in 12 mos. Reg BG bd. Gluc meter. 3/12 BG readings.
Tablets	Retain licence till 70	Retain licence unless diabetic eye affects VA.
Epilepsy	3 year licence for **6 months**	5- yrs fit-free and off rx
Faint (simple)	DVLA need not be notified	No driving restrictions

LOC   No driving restrictions Off for 3/12.
(unexplained syncope, but high probably vasovagal, ECG nad)

LOC with seizure   1 year refusal/ revoke   5 years refusal/ revoke
markers (tongue bite, incontinence, injury, confused)

LOC with no clinical pointers (after ix's)	Refuse/ revoke 6/12	Refuse/ revoke 1 year.
Heart failure	Need not notify	Disqualifies if symptomatic. Relicense if LVEF $\geq$ 0.4 and exercise test requirements met.
Hypertension	Need not notify	Disqualifies if SBP consistently $\geq$180 or and/or DBP $\geq$ 100 mm Hg.
Migraine	Not drive from onset of warning period.	
Psychosis	Stable for 3 months	Suspended licence for 3 years.
Vision (including cataract)	Read license plate at 20 metres.	Minimum 3/60 uncorrected or 6/9 corrected. Bar if > 6/9 in better eye or 6/12 in other.
Colour blind	Need not notify.	Need not notify.

FITNESS TO FLY - Contraindications
(IATA-BA Health Services A Guide to Physicians)

- **Respiratory** (SOB at rest, suspected PTX – wait 2/52 after drain) – must walk 50m flat or 1 flight of stairs
- **Heart disease** (unstable angina, poorly controlled heart failure, uncontrolled arrhythmia; < 10d post uncomplicated MI, 3-4 wks if complicated recovery); post-angio (3-5d)
- Stable **CVA** wait 3d but ideally 10d
- **DVT** before established on anticoagulants, can fly with LMWHep; untreated infectious disease
- **psychiatric illness** that could disrupt flight; **jaw** fractured with fixed wiring

- **Grand mal seizure** wait 24h; use colostomy bag if have colostomy (gas expansion)
- **Intraocular ops** and penetrating eye injuries – wait 1/52; no restrictions cataract or corneal laser
- **GI bleeding; anaemia** (<7.5 mg/dl); recent sickling crisis wait 10d;
- **> 36/40 nullip or > 32/40 multip;** hx of premature delivery, cervical incompetence; babies < 2 days old; need to carry cert after 28 weeks confirming EDD and that there are no complications.
- **< 10 days postop** hollow viscus or < 10 days post chest operation (CABG-allow air to be reabsorbed into chest)
- **Post orthopaedic cast** < 24h for < 2h flight or < 48h for longer flight. Exception bi-valved cast.

Advice: take insulin normal times, prophylactic ASA 75mg and stockings if at risk and ask airline leg room.

FORREST PLOT INTERPRETATION (sBMJ Vol 11 July 2003)

Each individual study is represented in the plot by 1 row. **A central rectangle** whose size is related to the number of patients in the study: bigger means more patients. The rectangle is centred on the overall result.

The **horizontal line** shows the 95% CIs for the study. A longer line means less certainty about the result or wide CIs. The **central vertical line** of the Forrest plot represents "no difference" between txs and distance along the horizontal axis represents increasing differences between the 2 txs.

The **horizontal axis** has a label telling you which tx is favoured to the left and which to the right.

At the end of several rows for individual studies is a **diamond** shape on the plot representing the meta-analysis summary of all the studies above it, again centred on the final result and showing the 95% CI.

Something else to look at on a Forrest plot is whether all the studies are consistent with each other. Whether the lines overlap each other is one way to assess this. A more formal method is to look for the **test for heterogeneity** statistic. If there is positive evidence that the studies are reporting different (heterogeneous) results, the P value will be significant. Hence a large P value, say > 0.1, reassures us that the studies are likely to be all measuring the same thing.

Local treatments for cutaneous warts: systematic review Sam Gibbs, Ian Harvey, Jane Sterling, Rosemary Stark BMJ 2002 325:461 31 Aug Fig 1 Cure rates in trials comparing topical salicylic acid with placebo for treatment of cutaneous warts

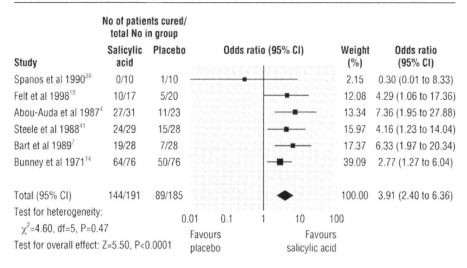

Study	No of patients cured/ total No in group		Odds ratio (95% CI)	Weight (%)	Odds ratio (95% CI)
	Salicylic acid	Placebo			
Spanos et al 1990[38]	0/10	1/10		2.15	0.30 (0.01 to 8.33)
Felt et al 1998[18]	10/17	5/20		12.08	4.29 (1.06 to 17.36)
Abou-Auda et al 1987[4]	27/31	11/23		13.34	7.36 (1.95 to 27.88)
Steele et al 1988[41]	24/29	15/28		15.97	4.16 (1.23 to 14.04)
Bart et al 1989[7]	19/28	7/28		17.37	6.33 (1.97 to 20.34)
Bunney et al 1971[14]	64/76	50/76		39.09	2.77 (1.27 to 6.04)
Total (95% CI)	144/191	89/185		100.00	3.91 (2.40 to 6.36)

Test for heterogeneity:
$\chi^2 = 4.60$, df=5, P=0.47
Test for overall effect: Z=5.50, P<0.0001

Fig 2 Cure rates in trials with cryotherapy and placebo or no treatment for treatment of cutaneous warts

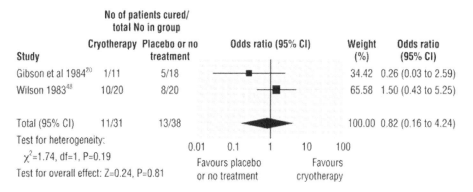

Study	No of patients cured/ total No in group		Odds ratio (95% CI)	Weight (%)	Odds ratio (95% CI)
	Cryotherapy	Placebo or no treatment			
Gibson et al 1984[20]	1/11	5/18		34.42	0.26 (0.03 to 2.59)
Wilson 1983[48]	10/20	8/20		65.58	1.50 (0.43 to 5.25)
Total (95% CI)	11/31	13/38		100.00	0.82 (0.16 to 4.24)

Test for heterogeneity:
$\chi^2 = 1.74$, df=1, P=0.19
Test for overall effect: Z=0.24, P=0.81

Fig 3 Cure rates in trials comparing cryotherapy with salicylic acid for treatment of cutaneous warts.

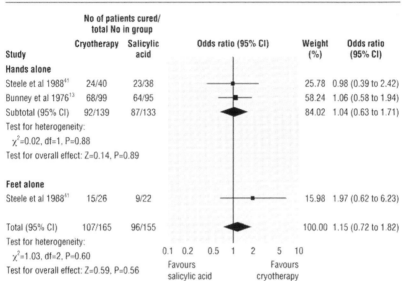

Components of a Forest plot The results of each trial in a systematic review are usually presented diagrammatically, comparing cure rates in trials of salicylic acid vs. placebo. This required explaining the meaning of the different components e.g. odds ratios, CI's, vertical line of no effect, squares, summation of the results and statistical significance.

Interpretation of the trials of salicylic acid vs. placebo

5 trials showed a statistically significant benefit of salicylic acid over placebo. However, the number of patients in each trial was small and the confidence intervals wide. One very small trial showed no difference between the 2 arms. The use of a log scale makes the results appear better. Aggregation makes them more meaningful.

Evaluation of the trials salicylic acid vs. placebo, cryo vs. placebo and cryo vs. salicylic acid

Salicylic acid was shown to be better than placebo and the result was statistically significant. 2 small trials showed no difference between cryotherapy and placebo. Also no difference was shown between cryotherapy and salicylic acid. Here the number of participants was larger. The results are confusing and do not help much in planning treatment. They do give evidence of the spontaneous resolution of warts in many patients. There is a lack of quality research and good randomised control trials are needed.

Patient factors influencing the management of warts

Given the poor quality of evidence, patient expectations fears and preferences are important. These will be influenced by health beliefs and the effects of treatment previously tried. Appearance, stigma, site, size and number of warts will also be important. The effects on work or leisure (e.g. swimming) will play a part. Young children will not tolerate cryotherapy. Immunocompromised patients may be treated more aggressively.

If in a trial the confidence interval crosses the vertical line of no effect then there is no statistically significant difference between the two arms of the trial. It is not appropriate to say that it shows a slight benefit in favour of one or other intervention. More critical appraisal of the likely quality of the individual trials making up the review would also enhance the answers. Comments about the numbers in the trials, the width of the confidence intervals and the weighting would help with this.

FUNNEL PLOT
(Scatter plot of treatment effect vs. study size)

Top of inverted funnel is narrow (more accurate) vs. bottom of funnel (more scattered – wide CI).

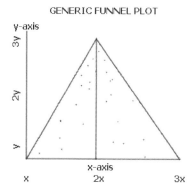

GENERIC FUNNEL PLOT

L'ABBE PLOT (circles correlate to sample size, line of infinity)

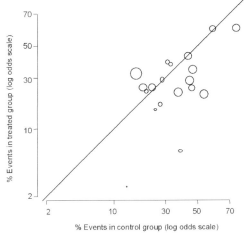

GLASGOW COMA SCALE - The GCS is scored between 3 and 15.

Best Eye Response. (4)
1.   No eye opening.
2.   Eye opening to pain.
3.   Eye opening to verbal command.
4.   Eyes open spontaneously.

Best Verbal Response. (5)
1.   No verbal response
2.   Incomprehensible sounds.
3.   Inappropriate words.
4.   Confused
5.   Orientated

Best Motor Response. (6)
1.   No motor response.
2.   Extension to pain.
3.   Flexion to pain.
4.   Withdrawal from pain.
5.   Localising pain.
6.   Obeys Commands.

A 'GCS of 11' is meaningless, break into its components, i.e. E3V3M5 = GCS 11. A Coma Score of ≥ 13 correlates with a mild brain injury, 9 to 12 moderate injury and ≤ 8 severe brain injury.

Teasdale G., Jennett B., LANCET (ii) 81-83, 1974.

GROIN LUMPS

Direct inguinal hernia - lump in superficial inguinal canal palpated on the pulp on the finger upon coughing;
- + cough impulse above inguinal ring;
- defect in posterior wall of the inguinal canal;
- medial to internal ring and inferior epigastric vessels;
- **above and medial to pubic tubercle**

Femoral aneurysm - mass with expansile pulsation in line of femoral artery

Femoral hernia
- most common in obese women
- below and lateral to pubic tubercle;
- absence of cough impulse above inguinal ring

Hydrocoele - can get above scrotal lump between thumb and finger and is translucent

Incarcerated inguinal hernia
- transmits no cough impulse
- tense, tender, irreducible

Indirect inguinal hernia
- lump in the superficial inguinal canal palpated on the tip of the finger upon coughing;
- + cough impulse above inguinal canal;
- patent process vaginalis;
- above and medial to pubic tubercle

Psoas Abscess
- retrofascial abscess
- may pass beneath the inguinal ligament and present in the upper part of the femoral triangle
- pain on hip extension or internal rotation
- x-ray shows loss of psoas shadow or mass
- soft, compressible, fluctuant mass
- may elicit fluctuation in parts of abscess above and below inguinal ligament

Saphena varix
- small lump medial to inguinal canal, disappears when lying + associated with varicosities
- soft compressible expansile dilatation at the top of the saphenous vein
- fluid thrill felt, if one percusses lower down the saphenous vein

HEPATITIS B SEROLOGY

	sAb	sAg	cAb	eAg	eAb	DNA
Immune, post vaccination	+ve	-ve	-ve	-ve	-ve	-ve
Immune, past exposure to HBV	+ve	-ve	+ve	-ve	+ve	-ve
Infected, low risk of transmission	–ve	+ve	+ve	-ve	+ve	-ve
Infected, high risk of transmission	–ve	+ve	+ve	+ve	-ve	+ve

HEPATITIS C: Dx to Tx (www.hepc.nhs.uk)

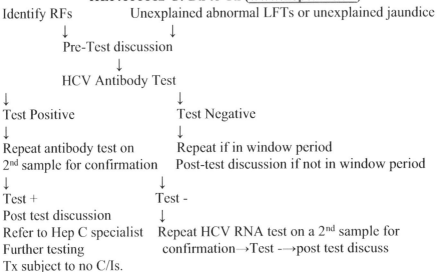

Identify RFs   Unexplained abnormal LFTs or unexplained jaundice
     ↓   ↓
     Pre-Test discussion
        ↓
     HCV Antibody Test
↓   ↓
Test Positive   Test Negative
↓   ↓
Repeat antibody test on Repeat if in window period
2nd sample for confirmation Post-test discussion if not in window period
↓   ↓
Test +   Test -
Post test discussion   ↓
Refer to Hep C specialist Repeat HCV RNA test on a 2nd sample for
Further testing   confirmation→Test -→post test discuss
Tx subject to no C/Is.
Tx: pegulated interferon and/or ribavarin 6-12/12.

HIV Transmission Rates

- **Mucocutaneous** transmission - 1 in 1000.
- **Needle-stick injury prophylaxis** – Explain that there is a 1 in 275 risk of acquiring HIV if exposure was to a known needle. Wash with soap and water and assess HIV risk. Prophylaxis consists of nelfinovir (5 tablets bd), AZT, and combivir to be taken for one month. This may also be offered for post-coital rape. Ideally PEP should be administered between 2h and 72h after exposure according to the CDC.
- **M to F** UPSI has an 8 x higher risk of transmission. **F to M** UPSI 1st encounter is 1-3%.

Management of Health and Safety at Work Regs 1999 Approved Code of Practice and Guidance

Reg 1 citation, commencement and interpretation comes into force in 1999

Reg 2 disapplication of these regs: master or crew of sea-giving ship, domestic service in a private household or work regulated as not harmful to young people in a family undertaking.

Reg 3 Risk assessment. Every employer shall make a suitable and sufficient assessment of-

A.   the risk to the health and safety of his employees to which they are exposed whilst they are at work;

B.   the risk to the health and safety of persons not in his employment arising out of or in connection with the conduct by him or his undertaking.

A hazard is something with the potential to cause harm.

A risk is the likelihood of potential harm from that hazard.

Ensure all aspects of the work activity are reviewed, including routine and non-routine activities.

Management of incidents and accidents.

Take account of risks to public.

Take account of need to cover fire risks (Part II of the Fire Regs).

Record significant findings of risk assessment if has 5 or more employees.

Review and revision.

Reg 4 Principles of prevention to be applied. Palliative measures. Adapt work Ensure that workers understand what they must do

Reg 5 Health and safety arrangements (planning, organisation, control, monitor, review). Adopt a systematic approach to the completion of a risk assessment.

Develop performance standards for the completion of the RA and the implementation of preventive and protective measures.

Reg 6 Health surveillance as required by COSHH. An identifiable disease or adverse health condition relate to the work concerned; and valid techniques are available to detect indications of the disease or condition;

and there is a reasonable likelihood that the disease or condition may occur under the particular conditions of work; and surveillance is likely to further the protection of the health and safety of the employees to be covered.

Health surveillance procedures

- inspection of readily detectable conditions by a responsible person
- enquiries about symptoms, inspection and exam by a qualified Occupational Health Nurse,
- medical surveillance (clinical exam, physiological or psychological effects)
- biological effect monitoring (i.e. diminished lung function in exposed workers),
- biological monitoring (measurement and assessment of the workplace agents or their metabolites either in tissues, secreta, excreta, expired air or any combination)

Reg 7 Health and Safety assistance (radiation protection adviser, occupational health advisor)

Reg 8 Procedures for serious and imminent danger and for danger areas

Reg 9 Contact with external services - procedures for any worker to follow in case of fire, for police or emergency service, for outbreak of public disorder.

Reg 10 Information for employees

Reg 11 Co-operation (with employer) and co-ordination (measures for his work)

Reg 13 Capabilities and training (demands of job should not exceed ability to carry out work)

Reg 14 Employee's duties (notify employer of any shortcomings and co-op with employer)

According to the National Disease Surveillance Centre, vomit should be cleaned with a dilute bleach solution (0.1% hypochlorite) to destroy the Norwalk-like virus (NLV) which is transmitted through the vomit of the infected.

HYPERTENSION MANAGEMENT
British Hypertension Society Guidelines 2004 (BHS-IV)
in line with European and WHO guidelines

Blood pressure	SBP (mm Hg)	DBP (mm Hg)
Optimal	< 120	< 80
Normal	< 130	< 85
High Normal	130-139	85-89
Grade 1 (mild HTN)	140-159	90-99
Grade 2 (moderate)	160-179	100-109
Grade 3 (severe)	≥180	≥110
Isolated systolic hypertension	140-159 (grade 1)	< 90
Isolated systolic hypertension	≥ 160 (grade 2)	< 90

BP (mm Hg)	Recommended action (advise all patients re lifestyle measures)
< 130/85	Reassess in 5 years
< 130-139/ 85-89	Reassess annually
140-159/90-99	Re-measure monthly if 10-year CHD risk < 20% and no target organ damage, no CV complications and no diabetes. Observe and reassess CHD risk annually.
140-159/ 90-99	If the 10-yr CHD risk is ≥ 20%, target organ damage, CV complications or diabetes, confirm over 12 weeks, then treat.
160-179/ 100-109	Re-measure weekly if 10-yr CHD risk is < 20% and no target organ damage, no CV complications and no diabetes. Treat if BP remains at this level over 4-12 weeks.
160-179/ 100-109	If the 10-yr CHD risk is ≥ 20%, target organ damage, CV complications or diabetes, confirm over 3-4 weeks, then treat if BP remains at this level.
≥ 160/100	Regardless of 10-yr CHD risk, confirm over 1-2 weeks, then treat.
> 180/ 110	Regardless of 10-yr CHD risk, confirm over 1-2 weeks, then treat. If malignant hypertension or hypertensive emergency, admit immediately.

Evaluation of hypertensive patients:
- Measure sitting BP every five years.
- Check standing BPs in elderly or diabetic patients at least at the initial estimation. Take mean of ≥ 2 BPs.
- Use ambulatory BP monitoring:
- Determining the efficacy of drug treatment over 24h
- Diagnosis and treatment of hypertension in pregnancy
- Evaluation of drug resistant hypertension
- Nocturnal hypertension
- Unusual variability of BP
- White coat hypertension
- Estimate 10-year CHD risk.
- Major risk factors include cardiovascular disease, coronary heart disease risk > 20%, diabetes and target organ damage. Other RFs include age, family history, male sex, TC: HDL ratio, and smoking.

Ix's: ECG, FBG, fasting lipids (TC, HDL, TG), creat/ lytes, urine dip for protein/ blood (**NOT CXR**).

Causes of hypertension:
- Coarctation (radiofemoral delay or weak femoral pulses)
- Conn's syndrome (hypokalaemia, muscle weakness, polyuria, tetany)
- Cushings disease
- Drugs (combined oral contraceptives, liquorice, NSAIDs, steroids, sympathomimetics)
- Pheochromocytoma (paroxysmal symptoms)
- Renal disease (proteinuria or haematuria, fhx, palpable kidney (hydronephrosis, neoplasm or polycystic)
- Renovascular disease (abdominal or loin bruit)

Target organ damage or complications:
- angina, CABG/ angioplasty, heart failure, LVH or left ventricular strain on ECG, MI
- carotid bruits, dementia, stroke, and transient ischaemia attack.
- fundal haemorrhages, papilloedema
- peripheral vascular disease
- proteinuria, renal impairment

Lifestyle measures: $\downarrow$ intake of total and saturated fat. $\downarrow$ salt (< 6g NaCl/day or 100 mmol/day). $\downarrow$ environmental stress and wt if BMI > 25 kg/m^2. 5 portions/day of fruit and veg. Stop smoking. Take up regular aerobic exercise for $\geq$ 30 mins/day at least 3 x a week. Limit weekly EtOH to

< 21 units (M), < 14 units (F).
- Add statin if CHD risk $\geq$ 20% and total cholesterol $\geq$ 3.5 mmol/l.
- Add aspirin 75 mg od if age $\geq$ 50 yo, with BP < 150/90 mm Hg and target organ damage, diabetes, or 10-yr CV risk $\geq$ 20%.

Refer to specialist:
- malignant hypertension/ emergency (accelerated hypertension with grade III-IV retinopathy)
- severe hypertension > 220/120 mm Hg
- impending complications of LVF, TIA, etc.
- hypertension in a patient < 20 yo or needing treatment in a patient < 30 yo
- investigation of underlying 2^0 causes
- resistance to multidrug regimen ($\geq$ 3 drugs)
- evaluation of therapeutic problems (multiple drug intolerance, contraindications, non-compliance)
- white coat hypertension, pregnancy and variable BP

DRUG TX OF HYPERTENSION - WHO GUIDELINES

Class of Drug	Indications	Contraindications
ACEI	CHF, LVF, post MI DM nephropathy	$\uparrow$ K, bilat renal art stenosis
α-blockers	BPH, glucose intolerance	orthostatic hypotension
Angiotensin II antag	ACEI cough	CHF, $\uparrow$ K, bilat renal art stenosis
β-blockers	after MI, angina tachyarrhythmias	CHF, asthma, COPD, heart block, DM
Ca antagonists	angina, elderly, systolic hypertension	grade 2 or 3 heart block with verapamil or diltiazem
Diuretics	CHF, elderly, systolic hypertension	gout

Mx OF HTN IN ADULTS IN 1° CARE (NICE Aug 2004)
- >140/90 - reassess BP from 2 further visits. Target BP is 140/90.
- Offer drug tx if >=160/100 OR Persistent BP > 140/90 + 10 yr CHD risk $\geq$ 15% OR CVD $\leq$ 20% OR existing CV disease or target organ
- Accelerated (malignant) HTN and suspected pheochromocytoma require immediate referral
- Offer patients > 80 yo and patients with isolated SBP, same tx as

patients with both raised SBP and DBP.

- Lifestyle advice - reduce caffeine, alcohol, diet, exercise, low dietary sodium, stop smoking
- Ix - urine for protein, blood (gluc, lytes, creat, TC, HDL chol), 12-lead ECG
- Refer - <30 yo, worsens suddenly, BP> 180/110 + signs of papilloedema   +/or retinal haem or responds poorly to tx.↑ creatinine (renal disease). Labile or postural hypotension, HA, palpitations, pallor, diaphoresis (pheo).↓K, abdo or flank bruits or ↑ creatinine when start ACEI, suggests renovascular hypertension. Cushing syndrome – osteoporosis, truncal   obesity, moon face, purple striae, muscle weakness, easy bruising, hirsutism, ↓K and hyperlipidaemia
- β-blockers may be considered in younger patients, particularly: those with an intolerance or CI to ACEI and AIIR antagonists, or women of child-bearing potential or patients with evidence of ↑ sympathetic drive.
- In patients whose BP is well-controlled (140/90 or <) with a regimen which includes a β-blocker, there is no absolute need to replace the β-blocker with an alternative agent.

Algorithm: tx of newly dx HTN (BHS, RCP, update to NICE 2004)

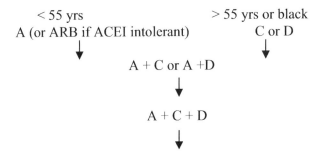

< 55 yrs   > 55 yrs or black
A (or ARB if ACEI intolerant)   C or D

A + C or A +D

A + C + D

Add: further diuretic treatment or selective α blocker or β blocker.
Consider seek specialist advice.

Mx OF BP AND LIPIDS IN TYPE 2 DIABETES (NICE Oct 2002)

- Annual estimate of heart disease risk with BP and blood lipid levels.
- If BP ≥ 140/80, and < 15% 10-year coronary event risk, then offer advice on lifestyle changes.
- If BP ≥ 140/80, and > 15% 10-year coronary event risk, then offer treatment.
- If BP ≥ 160/100, offer treatment to reduce BP to < 140/80.
- If BP ≥ 140/80 + albumin or protein in urine, then offer treatment to lower BP to ≤ 135/75.
- For people with a h/o CVD, give 75 mg aspirin a day.
- If TC level is ≥ 5.0 mmol/l or a TG level ≥ 2.3 mmol/l , then r/o 2^0 causes, control blood glucose levels, offer lifestyle advice, etc. Aim to ↓ TC to < 5 mmol/l or to 75-80% of the level prior to tx or to ↓ LDL-C to < 3 mmol/l or to 70% of the level prior to tx, whichever is lower.

MI: 2° PREVENTION IN 1° AND 2° CARE
(NICE May 07)

In the UK, approx 838,000 men and 394,000 women have had a MI at some point in their lives.

- After an acute MI, confirmation of the dx of acute MI and results of ix's, future mx plans and advice on 2° prevention should be part of every   discharge summary.
- Advise patients to undertake regular physical activity sufficient to ↑ exercise capacity.
- Advise patients to be physically active for 20–30 mins a day to the point of slight breathlessness. Patients who are not achieving this should be advised to ↑ their activity in a gradual, step-by-step way, aiming to ↑ their exercise capacity. They should start at a level that is comfortable, and ↑ the duration and intensity of activity as they gain fitness.
- Advise all patients who smoke to quit and offer assistance from a smoking cessation service 'Brief interventions and referral for smoking cessation in 1° care and other settings'
- Advise patients NOT to take supplements containing beta-carotene, and NOT to take antioxidant supplements (vitamin E and/or C) or folic acid to ↓ CV risk.

Diet: Advise patients to eat <u>min.7 g of omega 3 fatty acids per wk from 2 -4 portions of oily fish</u>.

For patients who have had an MI < 3 mos and who are not achieving 7 g of omega 3 fatty acids per week, consider providing at least 1 g daily of omega-3-acid ethyl esters tx licensed for 2° prevention post MI for up to 4 yrs.

- Advise patients to eat a <u>Mediterranean-style diet</u> (more bread, fruit, veg and fish; less meat; and replace butter and cheese with products based on veg and plant oils).
- Patients who drink <u>alcohol should be advised to keep weekly consumption within safe limits</u> (<21Us per week for M, or 14Us per week for F) and to avoid binge drinking (> 3 drinks in 1–2 hours).

After an MI, advise and support <u>all patients who are overweight or obese</u> to achieve and maintain a healthy wt in line with 'Obesity: NICE CG43.'

Cardiac rehabilitation should be equally accessible and relevant to all patients after an MI, particularly groups that are less likely to access this service (patients from black and minority ethnic groups, older patients, from lower socioeconomic groups, women, from rural communities and patients with mental and physical health comorbidities.) Programmes should include health ed and stress mx. A home based programme validated for patients who have had an MI (i.e. 'The Edinburgh heart manual'; see <u>www.cardiacrehabilitation.org.uk/heart_manual/heartmanual.htm</u>) that incorporates education, exercise and stress mx components with FUs by a trained facilitator may be used to provide cardiac rehab.

Most patients who have had an MI can return to work. Any advice should take into account the physical and psychological status of the pt, the nature of the work and the work environment.

Healthcare professionals should be up to date with the latest DVLA guidelines (<u>www.dvla.gov.uk</u>).

After an MI without complications, patients can usually travel by air within 2–3 weeks. Patients who have had a complicated MI need expert individual advice.

An estimate of the physical demand of a particular activity, and a comparison between activities, can be made using tables of metabolic equivalents (METS) of different activities (for further info please refer to <u>www.cdc.gov/nccdphp/dnpa/physical/measuring/met.htm</u>). Patients should also be advised how to use a perceived exertion scale to help monitor physiological demand. Patients who have had a complicated MI may need expert advice. Advice on competitive sport may need expert assessment of function and risk, and is dependent on what sport is being discussed and the level of competitiveness.

After recovery from an MI, **sexual activity** presents no greater risk of triggering a subsequent MI than if they had never had an MI. Patients who have made an uncomplicated recovery after their MI can resume sexual activity when they feel comfortable, **usually after 4 weeks**.

When treating erectile dysfunction, rx with a PDE5 (phosphodiesterase type 5) inhibitor may be considered in patients who had an MI > 6 mos earlier and who are now stable. PDE5 inhibitors must be avoided in patients treated with nitrates and/or nicorandil - can lead ↓↓BP.

Drug Treatment

Offer all patients who have had an acute MI a combination of: **ACE inhibitor, aspirin, b-blocker + statin.**

Early after presenting with an acute MI, offer all patients an ACEI initiated at the appropriate dose and titrated up at short intervals (i.e. q 1 to 2 wks) until the max tolerated or target dose is reached.

- <u>Assessment of LV function is recommended in all patients who have had an MI.</u> After an MI, all patients with preserved LV function or with LV systolic dysfunction should continue tx with an ACEI indefinitely, whether or not they have sxs of heart failure.
- Routine rx of angiotensin receptor blockers (ARBs) after an acute MI is NOT recommended. Offer an ARB for those who have had to discontinue an ACEI due to intolerance (i.e. cough) or allergy.
- Combined treatment with an ACEI and an ARB is NOT recommended for routine use in patients early after an acute MI with heart failure and/or LV systolic dysfunction.
- In patients with a proven MI in the past (> 1 yr ago) and with heart failure and LV systolic dysfunction, ACEI and ARB rx should be in line with 'Chronic heart failure' (NICE CG 5).
- In patients with a proven MI in the past and with LV systolic dysfunction, who are asymptomatic, ACEI should be offered and the dose titrated upwards, as tolerated, to the effective clinical dose for patients with heart failure and LV systolic dysfunction.
- In patients with a proven MI in the past without heart failure and with preserved LV function, ACEI should be offered and the dose titrated up, as tolerated, to the effective clinical dose.

- In patients with a proven MI in the past with LV systolic dysfunction, who are asymptomatic and who have had to discontinue an ACEI because of intolerance, an ARB should be substituted.

Measure renal function, serum lytes and BP before starting an ACEI or ARB and again w/n 1 or 2 wks of starting tx. Monitor patients as appropriate as the dose is titrated up, until the max tolerated or target dose is reached, and then at least annually. Monitor patients with CHF in line with NICE.

Antiplatelet therapy
Aspirin should be offered to all patients after an MI, and should be continued indefinitely.

- Clopidogrel should NOT be offered as 1ˢᵗ-line monotherapy after an MI. Clopidogrel, in combination with low-dose aspirin, is recommended for use in the mx of non-ST-segment-elevation acute coronary syndrome in patients who are at moderate to high risk of MI or death.
Patients at moderate to high risk of MI or death, presenting with non-ST-segment-elevation acute coronary syndrome can be determined by clinical signs and sxs, accompanied by 1 or both of the following:
- the results of clinical ix's, i.e. new ECG changes (other than persistent ST segment elevation) indicating ongoing myocardial ischaemia, particularly   dynamic or unstable patterns
- the presence of ↑ blood levels of markers of cardiac cell damage such as troponin.
- Tx with clopidogrel in combination with low-dose aspirin should be continued for 12 mos after the most recent acute episode of non-ST-segment-elevation acute coronary syndrome. Thereafter, standard care, including tx with low-dose aspirin alone, is recommended unless there are other indications to continue dual antiplatelet tx.

Renal function and serum K should be monitored before and during tx with an aldosterone antagonist. If hyperkalaemia is a problem, the dose of the aldosterone antag should be halved or stopped.

Statins and other lipid lowering agents
Statin therapy is recommended for adults with clinical evidence of CV disease in line with 'Statins for the prevention of cardiovascular events' (NICE technology appraisal guidance 94)

After an MI, all patients should be offered treatment with a statin ASAP.
The decision whether to initiate statin tx should be made after an informed discussion between the healthcare professional and the individual about the risks and benefits of statin tx, and taking into account additional factors such as comorbidities and life expectancy.

Baseline liver enzymes should be measured before initiation of a statin.
Patients who have ↑d liver enzymes should NOT routinely be excluded from statin therapy.
When the decision has been made to prescribe a statin, the treatment should usually be initiated with a drug with a low acquisition cost (taking into account required daily dose and product price per dose).
Patients who are intolerant of statins should be considered for other lipid lowering agents.

- Routine monitoring of CK in asymptomatic patients who are treated with a statin after an MI is NOT recommended.
- Patients who are being treated with a statin and who develop muscle sxs (pain, tenderness or weakness) should be advised to seek medical advice so that creatinine kinase can be measured.
  The dose of any statin may need to be reduced or stopped if there are issues surrounding the metabolic pathway, food and/or drug interactions and/or concomitant illness.
  Statins should be discontinued in patients who develop peripheral neuropathy that may be attributable to the statin tx, and further advice from a specialist should be sought.
- After an ST-segment-elevation MI, patients treated with a combination of aspirin and clopidogrel during the first 24 hrs after the MI should continue this treatment for at least 4 wks. Thereafter, standard treatment including low-dose aspirin should be given, unless there are other indications to continue dual antiplatelet treatment.
- If the patient has not been treated with a combination of aspirin and clopidogrel during the acute phase of an MI, this combination should not routinely be initiated.
  The combination of aspirin and clopidogrel is NOT recommended for routine use for > 12 months after the acute phase of MI, unless there are other indications to continue dual antiplatelet treatment, and the combination is usually recommended for a shorter duration after an ST-segment-elevation MI.

For patients with aspirin hypersensitivity, clopidogrel monotherapy should be considered as an alternative tx

In patients with a history of dyspepsia, tx with a proton pump inhibitor and low-dose aspirin.

After appropriate tx, patients with a hx of aspirin-induced ulcer bleeding whose ulcers have healed and who are negative for *Helicobacter pylori* should be considered for tx with a full-dose PPI and low-dose aspirin.

β-blockers - Early after an acute MI, all patients without LV systolic dysfunction or with LV systolic dysfunction (symptomatic or asymptomatic) should be offered tx with a b-blocker.

β-blockers should be initiated ASAP when the pt is clinically stable and titrated upwards to the max tolerated dose. B-blockers should be continued indefinitely after an acute MI.

For patients after an MI with LV systolic dysfunction, who are being offered tx with a beta-blocker, clinicians may prefer to consider tx with a beta-blocker licensed for use in heart failure.

After a proven MI in the past, all patients with LV systolic dysfunction should be offered tx with a b-blocker whether or not they have sxs, and those with heart failure plus LV systolic dysfunction should be managed in line with 'Chronic heart failure' (NICE CG 5).

After a proven MI in the past, patients with preserved LV function who are asymptomatic should NOT be routinely offered tx with a β-blocker, unless they are identified to be at ↑ed risk of further CV events, or there are other compelling indications for b-blocker tx.

Vitamin K antagonists
For patients who have had an MI, high-intensity warfarin (INR >3) should NOT be considered as an alternative to aspirin in 1st-line treatment.

For patients who have had an MI and are unable to tolerate either aspirin or clopidogrel, treatment with moderate-intensity warfarin (INR 2–3) should be considered for up to 4 years, and possibly longer.

For patients who have had an acute MI, are intolerant to clopidogrel and have a low risk of bleeding, treatment with aspirin and moderate-intensity warfarin (INR 2–3) combined should be considered.

For patients already being treated for another indication (mechanical valve, recurrent DVT, atrial fib, LV thrombus), warfarin should be continued. For patients treated with moderate-intensity warfarin (INR 2–3) and who are at low risk of bleeding, consider the addition of aspirin.

The combination of warfarin and clopidogrel is NOT routinely recommended.

Ca channel blockers should NOT routinely be used to ↓ CV risk after an MI.
If b-blockers are C/I or need to be discontinued, diltiazem or verapamil may be considered for 2° prevention in patients without pulmonary congestion or LV systolic dysfunction.
For patients who are stable after an MI, calcium channel blockers may be used to treat hypertension and/or angina. For patients with heart failure, amlodipine should be used, and verapamil, diltiazem and short-acting dihydropyridine agents should be avoided in line with 'Chronic heart failure' (NICE CG 5).
K channel activators   Nicorandil is NOT recommended to ↓ CV risk in patients after an MI.

Aldosterone antagonists in patients with heart failure and LV dysfunction
For patients who have had an acute MI and who have sxs and/or signs of heart failure and LV systolic dysfunction, tx with an aldosterone antagonist licensed for post-MI tx should be initiated w/n 3–14 days of the MI, preferably after ACEI.
Patients who have recently had an acute MI and have clinical heart failure and LV systolic dysfunction, but who are already being treated with an aldosterone antagonist for a concomitant condition (i.e., CHF), should continue with aldosterone antagonist or alternative, licensed for early post-MI tx
For patients who have had a proven MI in the past and heart failure due to LV systolic dysfunction, tx with an aldosterone antagonist should be in line with NICE: CHF.
Offer all patients a cardiological assessment to consider whether coronary revascularisation is appropriate. Take into account comorbidity.

Patients with hypertension Hypertension should be treated to the currently recommended target of ≤140/90 mmHg given in NICE Patients with relevant comorbidities, DM or renal disease, treat to a lower BP.

Consider an implantable cardioverter defibrillator in patients with LV systolic dysfunction in line with 'Implantable cardioverter defibrillators for arrhythmias' (NICE 95).

Chest Pain (NICE Mar 2010)

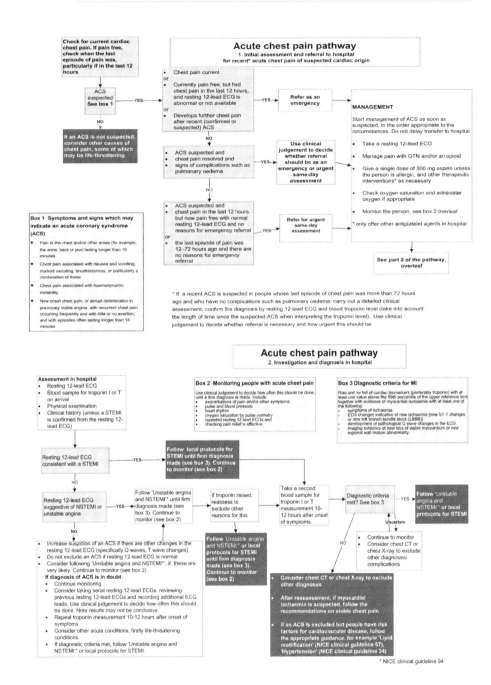

Stable chest pain pathway
1. Presentation

Carry out a detailed assessment and review
History
Document:
- the age and sex of the person
- the characteristics of the pain and any associated symptoms
- any history of angina, MI, coronary revascularisation, or other cardiovascular disease and
- any cardiovascular risk factors.

Examination
- Identify risk factors and signs of cardiovascular disease
- Identify non-coronary causes of angina (for example, severe aortic stenosis, cardiomyopathy)
- Exclude other causes of chest pain

Box 1 Typical stable angina symptoms
- Constricting discomfort in the front of the chest, in the neck, shoulders, jaw, or arms
- Precipitated by physical exertion
- Relieved by rest or GTN within about 5 minutes

Typical angina: all of the above
Atypical angina: two of the above
Non-anginal chest pain: one or none of the above

See recommendation 1.3.3.4 for risk factors which make angina more likely.

- Consider other causes of chest pain
- Only consider chest X-ray if other diagnoses are suspected

◄YES

- Features of pain are non-anginal (see boxes 1 and 2) and
- Assessment does not raise clinical suspicion of stable angina

Person has confirmed CAD

YES

See part 3 of the pathway on page 52

Box 2
Stable angina is unlikely if chest pain is:
- continuous or very prolonged and/or
- unrelated to activity and/or
- brought on by breathing in and/or
- associated with symptoms such as dizziness, palpitations, tingling or difficulty swallowing

NO

Take resting 12-lead ECG (see box 3)

- Consider other causes of chest pain
- Consider investigating other causes of angina such as hypertrophic cardiomyopathy in people with typical angina-like chest pain and a low likelihood of CAD (< 10%)
- Only consider chest X-ray if other diagnoses are suspected

Likelihood of CAD is less than 10%

Use clinical assessment and typicality of anginal pain features to stratify the likelihood of CAD (see box 1 and table 1)

Likelihood of CAD is greater than 90%

to identify conditions which exacerbate angina
Treat as stable angina

Likelihood of CAD is 10- 90%

- Arrange blood tests to identify conditions which exacerbate angina
- Offer further diagnostic testing (see part 2 of pathway on page 51)
- Consider aspirin only if the chest pain is likely to be stable angina until diagnosis made
- Follow local protocols for stable angina while waiting for the results of investigations if symptoms are typical of stable angina.

Box 3 Changes on a resting 12-lead ECG consistent with CAD which may indicate ischaemia or previous infarction
- pathological Q waves in particular
- LBBB
- ST-segment and T wave abnormalities (for example, flattening or inversion).

Results may not be conclusive. Consider resting 12-lead ECG changes together with people's clinical history and risk factors. Note that a normal resting 12-lead ECG does not rule out stable angina.

Stable chest pain pathway
2. Diagnostic testing for people in whom stable angina cannot be diagnosed or excluded by clinical assessment alone

Estimated likelihood of CAD 10 to 29%

CT calcium scoring

score is zero

score is more than 400 → Follow pathway for 61-90% CAD

score is 1- 400

64-slice (or above) CT coronary angiography

Investigate other causes of chest pain**

◄NO

Significant CAD See box 4

Uncertain

YES

Treat as stable angina

Appropriate functional imaging test (see box 5 overleaf). If reversible myocardial ischaemia found, treat as stable angina. If not, investigate other causes of chest pain**

Estimated likelihood of CAD 30-60%

Appropriate functional imaging test (see box 5 overleaf)

Reversible myocardial ischaemia

Investigate other causes of chest pain**

NO

Uncertain

Yes

Treat as stable angina

Invasive coronary angiography

Investigate other causes of chest pain**

◄NO

Significant CAD See box 4

YES

Treat as stable angina

Estimated likelihood of CAD 61-90%

Invasive coronary angiography if appropriate*

Treat as stable angina

◄YES

Significant CAD See box 4

NO►

Investigate other causes of chest pain **

Uncertain

Appropriate functional imaging test (see box 5 overleaf)

Reversible myocardial ischaemia

NO►

Investigate other causes of chest pain **

YES

Treat as stable angina

Box 4 Definition of significant coronary artery disease

Significant coronary artery disease (CAD) found during invasive coronary angiography is ≥ 70% diameter stenosis of at least one major epicardial artery segment or ≥50% diameter stenosis in the left main coronary artery.

a) Factors intensifying ischaemia. Such factors allow less severe lesions (for example ≥50%) to produce angina.
- Reduced oxygen delivery: anaemia, coronary spasm
- Increased oxygen demand: tachycardia, left ventricular hypertrophy
- Large mass of ischaemic myocardium: proximally located lesions
- Longer lesion length

b) Factors reducing ischaemia. Such factors may render severe lesions (≥70%) asymptomatic.
- Well developed collateral supply
- Small mass of ischaemic myocardium: distally located lesions, old infarction in the territory of coronary supply.

* If coronary revascularisation is not being considered or invasive coronary angiography is not appropriate or acceptable to the person, offer non-invasive functional imaging

**Consider investigating other causes of angina, such as hypertrophic cardiomyopathy or syndrome X in people with typical angina-like chest pain if investigation excludes flow-limiting disease in the epicardial coronary arteries.

IRRITABLE BOWEL SYNDROME IN ADULTS (NICE Feb 08)
Person with any of these sxs for at least 6/12
(Abdo pain/discomfort, Bloating, Change in bowel habit)

↓

Patient hx and clin exam → **Red Flag sxs** (Rectal bleeding,
by GP/ 1°care clinician Unexplained unintentional weight loss,
IBS Positive Dx Criteria fhx of bowel/ovarian CA, late onset
Ixs in PC (FBC anaemia, ESR, 60 yo. Assess for anaemia, abdo,
CRP – inflammatory bowel disease, pelvic (if appropriate), rectal masses
EMA or TTG coeliac) and inflammatory bowel disease.
Immediate referral to 2° care.

↓

IBS Mx

Base on nature and severity of sxs and
individual or combo of meds, lifestyle advice,
directed at the main sx(s).

Lifestyle: Diet & Physical Activity   **Drug Tx** consider single or combo:
Assess diet:↓fibre intake; take soluble fibre;   Antispasmodics
consider dietitian referral
Assess level of phys activity: encourage ↑   Antimotility agents (titrate dose)
Pt info resource with dietary, lifestyle   Laxatives (titrate dose)
and self help advice   **2nd line TCA (or SSRIs)**
Effective sx control. FU to evaluate response (timescale negotiated between doctor and patient). Not effective. Continuing sx profile. >12 months' duration, consider **psychological interventions** (**hypnotherapy**, psychological tx, CBT).

LONG-ACTING REVERSIBLE CONTRACEPTION (NICE Oct 05)
30% pregnancies unplanned. Uptake of LARC 8% 18-49 yo 2003-04 vs. 25% ocp vs. 23% condom.
LARC - copper IUD, progest-only (IUS, injectables, subdermal implants) or combined vaginal rings.
More cost effective than coc at 1 yr of use. IUDs, IUS and implants more cost effective vs. injectables.
Counselling: efficacy, duration of use, risk and possible S/Es, non-contraceptive benefits, procedure for initiation and removal/ discontinuation, when to seek help while using the method
Contraceptive prescribing - med hx, fhx, menstrual, contraceptive and sexual hx. Exclude pregnancy. Supply interim method of contraception at 1st

appt if required; obtain informed consent, refer VTE to contraception specialist. Promote safe sex. Assess risk for STIs and advise testing when appropriate. Info on local STI screening. Fraser guidelines for < 16 yos. Contraception in terms of needs of individual rather than relieving anxieties of carers or relatives. If learning disability, carers and parties involved should establish a care plan.

IUD/ IUS only fitted by trained person with continuous experience of inserting min. one IUD or IUS a mo.

Copper IUD

- IUDs prevent fertilisation and inhibit implantation. The licensed duration of use containing 380 mm^2 copper ranges from 5-10 yrs. Pregnancy rate is < 20 in 1000 over 5 yrs. No evidence of delay in return of fertility following removal or expulsion. Heavier bleeding +/or dysmenorrhoea are likely.
- Up to 50% stop using IUDs within 5 years due to unacceptable PVB and pain. Does not affect weight.
- Risk of uterine perforation at time of insertion < 1/1000. Risk of PID < 1/100. Risk of expulsion < 1/20 in 5 yrs. Overall risk of ectopic 1/1000 in 5 years. If pregnancy with IUD in situ, risk of ectopic is 1/20.
- ≥ 40 yo may retain device until no longer require contraception.
- Test for chlamydia, N gonorrhoea, and any STIs in women who request before IUD insertion. If not possible, then give prophylactic antibiotics before IUD insertion in those at ↑ risk of STIs.
- May use in adolescents but consider STI risk. Not C/I in nullips of any age. Not C/I in diabetes.
- Safe to used when breastfeeding and for HIV + or with AIDS.
- May insert at any time during cycle, immediately post 1st or 2nd trim abortion or from 4 wks postpartum.
- Emergency drugs (AEDs) should be available at time of insertion in epileptics as ↑ risk of seizure at time of cervical dilation.
- Check threads, light bleed for few days and pain for few hours, see sooner if symptoms of perforation or infection.
- F/U after 1st menses or 3-6 weeks after insertion to rule out infection, perforation or expulsion.
- Tx heavier periods with NSAIDs and tranexamic acid or if unacceptable, change to LNG-IUS.
- Actinomyces organisms on smear with IUD requires r/o PID. Routine removal not indicated.

- Remove IUD < 12 wks gestation if have IUP.

IUS

- Acts predominantly by preventing implantation and sometimes fertilisation.
- Pregnancy rate is < 10 in 1000 over 5 yrs. Licensed use for 5 years duration. No delay in return of fertility.
- Irregular bleed and spot for 1st 6 mos. Oligomenorrhoea or amenorrhoea by the end of the 1st yr.
- 60% stop IUS within 5 yrs due to unacceptable PVB and pain. No weight gain. ↑ acne as a result of absorption of progestogen. Risk of uterine perforation at   insertion < 1 in 1000. Risk of PID < 1 in 100. Expulsion < 1 in 20 in 5 yrs. Risk of ectopic < than no contraception. Overall ectopic risk is 1 in 1000 in 5 yrs. Ectopic 1 in 20 if pregnant.
- ≥ 45 yo may retain IUS until they no longer require contraception. STI testing before insertion. May be used in adolescents but consider STI risk; not C/I in nullips and IS safe in breastfeeding. Is safe to use if oestrogen is C/I. Safe for women with HIV or AIDs. Not C/I in diabetics.
- Insert at any time during cycle (but if amenorrhoeic or it has been > 5d since menstrual bleeding started, additional barrier contraception should be used for the 1st 7 days after insertion); immediately after 1st or 2nd trimester abortion or any time thereafter; from 4 wks postpartum.
- Emergency drugs including AED to be made available at time of IUS insertion in epileptics.
- F/U after 1st menses or 3-6 weeks after insertion.

Progestogen-only injectables

- Act by preventing ovulation. Pregnancy rate is < 4 in 1000 over 2 years. Rate of DMPA is < NET-EN.
- DMPA (depot) q 12/52 and NET-EN (norethisterone enantate) q 8/52. Delay of fertility up to 1 year.
- If she stops, should use a different contraceptive method immediately, even if amenorrhoea persists.
- Amenorrhoea is more likely with DMPA than NET-EN, as time goes by and is not harmful.
- DMPA may be associated with weight ↑ of up to 2-3 kg > 1 year. Is NOT associated with acne, depression, HA's.

- DMPA is associated with small loss of BMD, which largely recovers when stopped. No evidence of $\uparrow$ fracture.
- May be given in adolescents or > 40 yo if other methods are not acceptable.
- BMI > 30 can safely use DMPA and NET-EN and breastfeeding is safe.
- May be used by women who have migraines +/- aura. Medically safe if oestrogen is C/I.
- May be associated with $\downarrow$ in seizure freq in epileptic. No evidence of $\uparrow$ `risk of STI or HIV. Safe +HIV/ AIDS.
- May use if taking liver enzyme-inducing meds and the dose interval does not have to be reduced.
- IM into gluteal or deltoid or lateral thigh.
- Start up to and including the 5th day of the cycle without the need for additional cover. At any other time, need to use additional barrier for 1st 7 days after injection; may be used immediately after 1st or 2nd trimester abortion or any time thereafter or at any time postpartum.
- May give if up to 2 weeks late for repeat DMPA without need for additional contraception.
- Persistent PVB associated with DMPA treatment with mefenamic acid or ethinyestradiol.
- Review if wish to continue > 2 years. No evidence of congenital malformations if pregnant occurs on DMPA.

Progestogen-only subdermal implants (implanon)

- Prevents ovulation. Pregnancy rate < 1 in 1000 over 3 years. Use for 3 years. No delay in return of fertility.
- Bleeding patterns likely to change. 20% have no bleed, 50% have infrequent, frequent or prolonged PVB.
- Likely to remain irregular over time. Dysmenorrhoea may be $\downarrow$ during the use of implanon.
- Up to 43% stop using implanon w/n 3 yrs due to 33% irregular bleed and $< 10\%$ hormonal problems.
- Not associated with weight change, mood, libido or HA's. May be associated with acne.
- > 70 kg can use. Breastfeeding safe. Not C/I in diabetes. No $\uparrow$ risk of STI or HIV. Safe for HIV + or AIDs. May be used in patients with migraine +/- aura. Medically safe if oestrogen is C/I; no evidence of effect on BMD.

- Implanon is NOT recommended for women taking liver enzyme-inducing drugs.
- Insertion: at any time (but if amenorrhoeic or > 5days since menses started, additional barrier for 1ˢᵗ 7 days; immediately after abortion and at any time post-partum.
- No routine F/U. Tx irregular bleed with ponstan, ethinyestradiol or mifepristone. No teratogenic effect.
- If cannot palpate implanon (deep or failed insertion or migration), localise by USG before remove.

LUNG CANCER - The diagnosis and treatment (NICE Feb 2005)
Lung CA #1 cause of CA death in men and #2 in women (breast #1).
21% alive after 1 year, 5% - 5 years.
Types - small-cell lung CA (20%) vs. NSCLC (80% - 35% SqCCa, 27% adenoCA, 10% large cell CA)

Urgent referral for CXR:
Haemoptysis OR Any of the following unexplained or persistent (>3/52) sxs/ signs:

- Cough
- Chest/ shoulder pain
- Dyspnoea
- Weight loss
- Features suggestive of met from lung CA (in brain, bone, skin)

Chest signs
Hoarseness
Finger clubbing
Cervical / SC LNs

Urgent referral to a chest physician while awaiting results of CXR:
- a CXR or CT scan suggests lung CA (pleural effusion and slowly resolving consolidation)
- persistent haemoptysis in smokers/ ex-smokers > 40 yo
- signs of SVC obstruction (swelling of face/ neck with fixed ↑ of JVP
- stridor

Diagnosis
- CXR suggestive of lung CA, send second copy to chest physician and offer a contrast-enhanced CT scan including liver and adrenals to further diagnosis and stage the disease
- Chest CT before fibreoptic bronchoscopy or any other bx procedure
- Bronchoscopy for patients with central lesions

- Sputum cytology is rarely indicated and reserved for centrally placed nodules or masses and are unable to tolerate bronchoscopy or other invasive tests
- Percutaneous transthoracic needle bx for peripheral lesions
- Surgical biopsy where other less invasive methods of bx fail. Biopsies from metastatic site if achieved easily than from 1°.
- FDG-PET to investigate solitary pulmonary nodules in cases where a bx is not possible or has failed. Every cancer network should have rapid access to FDG-PET scanning

Staging for NSCLC:

- CT alone may not be reliable in assessment of mediastinal and CW invasion - U/S or surgical assessment
- MRI should not routinely be performed to assess the stage of the primary tumour
- MRI to assess the extent of disease for patients with superior sulcus tumours; clinical signs/sxs of brain mets
- FDG-PET scan for: candidates staged for surgery on CT to look for involved intrathoracic LNs and distant mets; surgical candidates with N2/3 disease; candidates for radical radiotx.
- Histological/ cytological investigation - to confirm N2/3 disease where FDG-PET is positive.
- Histological/ cytological confirmation of LNs - should not be performed before surgical resection for N0 or N1 and M0 (stage 1 and II) by CT and FDG-PET; where there is definite distant met disease, where there is a high probability that the N2/N3 disease is metastatic (chain of high FDG uptake in LNs)
- Biopsy is not required when FDG-PET scan for N2/N3 disease is negative even if LNs are enlarged on CT
- If FDG-PET is not available, suspected N2/3 disease by CT scan should be histologically sampled in surgical or radical radiotx candidates
- X-ray for localised signs or sxs of bone mets. If inconclusive or negative, either a bone scan or MRI

Treatment for NSCLC:
Surgery with curative intent - FU for 9 mos after and not > 5yrs
- 1st line for Stage I or II, lobectomy OR limited resection or radical radiotx if cannot tolerate

- Sleeve lobectomy alternative to pneumonectomy for stage 1, II with central tumour to conserve functioning lung
- T3 with CW involvement - complete resection of the tumour by extrapleural or en bloc CW resection
- All patients undergoing surgical resection should have systematic LN sampling for pathological staging
- Stage IIIa (N2) - poor prognosis for surgery alone. **Radical radiotherapy alone** - FU for 9 months and not > 5 years
- Stage I, II or III and should undergo pulmonary function tests before radiotherapy and if poor, offer radiotherapy if volume of irradiated lung is small.
- Stage I or II who are medically inoperable but suitable for radiotx should be offered continuous hyperfractionated accelerated radiotherapy (CHART).
- Stage IIIA or IIIB who cannot tolerate chemotherapy should be offered CHART.
- If CHART is not available, conventional fractionated radiotherapy in 20 fractions over 4 weeks.

Chemotherapy - 1 month follow-up with CXR

- Stage III or IV
- Advanced NSCLC - combo of single 3rd gen drug (docetaxel, gemcitabine, paclitaxel or vinorelbine) + a platinum drug (either carboplatin or cisplatin). If unable to tolerate a platinum combo then offer a single-agent chemotx with a 3rd generation drug.
- Docextaxel monotherapy if 2nd line tx is appropriate for patients with locally advanced or met NSCLC in whom relapse has occurred after previous chemo.

Combination therapy

- Stage I, II or IIIA who are suitable for resection should NOT be offered preop chemo
- Preop radiotx is NOT recommended for surgical candidates
- Postop radiotx is NOT recommended for patients with NSCLC after complete resection but SHOULD be considered after incomplete resection of the primary tumour
- Adjuvant chemo should be offered who have had a complete resection

- Pathologically staged II and III following resection should NOT receive postop chemo
- Stage III who are not suitable for surgery but eligible for radiotx should be offered sequential chemo

Staging for SCLC
- By contrast-enhanced CT scan of chest, liver, adrenals + selected imaging of any symptomatic area

Treatment for SCLC
- Assessment of major prognostic factors: performance status, serum lactate dehydrogenase, liver (LFTs), serum sodium and stage
- All patients should be offered platinum chemo and multidrug regimens
- 4-6 cycles of chemo should be offered to patients whose disease responds. No maintenance tx.
- limited-stage SCLC - offer thoracic irradiation concurrently with 1st or 2nd cycle chemo or following completion of chemo if had good partial response w/n the thorax. Extensive disease, thoracic irradiation following chemo if had a complete response at distant sites and a good partial response w/n the thorax
- Consolidation thoracic irradiation - 25 fractions over 5 wks
- Prophylactic cranial irradiation - limited disease and complete or good partial response after primary tx
- 2nd-line chemo to patients at relapse only if their disease responded to 1st-line chemo

Palliative - breathlessness clinic, OT, PT, 1 mo FU with CXR after palliative radiotherapy
- external beam radiotx for relief of breathlessness, cough, haemoptysis or chest pain
- opioids, ex codeine or morphine, to reduce cough
- debulking bronchoscopic procedures for relief of distressing large-airway obstruction or bleeding due to an endobronchial tumour within a large airway
- extrinsic compression consider stents
- troublesome hoarseness due to recurrent laryngeal N palsy – refer ENT
- SVC obstruction - offer chemo and radiotx, consider stent for immediate relief

- Cerebral mets - corticosteroids and radiotx
- Other sxs - wt loss, loss of appetite, depression, difficulty swallowing - lung MDT
- Pleural aspiration or drainage; talc pleurodesis for long-term benefit
- Bone met - standard analgaesia - single-fraction radiotx if not adequate
- Spinal cord compression - medical emergency and tx < 24h with corticosteroids, radiotx and surgery and early referral to oncology physiotherapist and OT for assessment, tx and rehab

MEDICAL CERTIFICATES
www.dwp.gov.uk

Self-cert (SC1)   Patients who are not eligible for statutory sick pay but who wish to claim for incapacity benefit.
Certifies for first 7 days of illness. Covers NI contributions for the self-employed.

Self-cert(SC2)   as SC1 but for statutory sick pay. Forms from the surgery, boss or local Jobcentre Plus and covers 1st 7 days off work.

Med 10   hospital in-patient and also consider med 3 on discharge for future recovery period.

Revised Med 3 (statement of fitness to work or fit note) replaces med 3 and med 5, may be issued for a maximum period of 3 months for a period of anticipated incapacity ie had treatment in A+E and unable to work for 7 days. Does not need to return to you for a new statement when fit to work.

Med4   after 28w incapacity; patient is sent form IB50 and asked for med4 from GP. Dept of Works and Pensions may ask for further medical exam. Personal capability test (All Works Test).

Med6   a vague dx is put on the form if deemed harmful to the patient or detrimental for the employer to know the true dx; the med6 requests the Jobcentre to send an additional form for more details.

RM7   form sent to Jobcentre Plus asking them to review patient sooner if doubt disability.

DS1500   Terminal illness, not expected to live > 6/12.

HEAVY MENSTRUAL BLEEDING (NICE Jan 2007)

For clinical purposes, HMB defined as excessive menstrual blood loss which interferes with the woman's physical, emotional, social and material quality of life, and which can occur alone or in combination with other sxs. Any interventions should aim to improve quality of life measures.

Hx taking, exam and ixs

- If the hx suggests HMB without structural or histological abnormality, pharmaceutical tx can be started without carrying out a physical exam or other ix's at initial consultation in primary care, unless the tx chosen is levonorgestrel-releasing intrauterine system (LNG-IUS).
- If the hx suggests HMB with structural or histological abnormality, with sxs such as IMB or postcoital bleed, pelvic pain +/or pressure sxs, perform a phys exam and/or other ix's (scan).
- Carry out exam before all: LNG-IUS fittings, ix's for structural and for histological abnormalities.
- Women with fibroids that are palpable abdominally or who have intracavity fibroids and/or whose uterine length as measured at scan or hysteroscopy is > 12 cm refer immediately to a specialist.

Ix's **FBC** should be carried out on all. This should be done in parallel with any HMB tx offered.
- Consider test for coagulation disorders (i.e., von Willebrand's disease) in women who have had HMB since menarche and have personal or family hx suggesting a coagulation disorder.
- A serum ferritin test should NOT routinely be carried out for HMB.
- Female hormone testing should NOT be carried out for HMB.
- **TFT's should be carried out ONLY when other signs/ sxs of thyroid disease are present**.
- If appropriate, a biopsy should be taken to exclude endometrial CA or atypical hyperplasia. Indications for a biopsy: persistent IMB, and in women aged ≥ 45 yo, tx failure or ineffective tx.
- Imaging should be undertaken in the following circumstances: The uterus is palpable abdominally. Vaginal exam reveals a pelvic mass of uncertain origin. Pharmaceutical tx fails.
- **Ultrasound is the 1st-line diagnostic tool** for identifying structural abnormalities.

- Hysteroscopy should be used as a diagnostic tool only when U/S results are inconclusive, i.e., to determine the exact location of a fibroid or the exact nature of the abnormality.
- If imaging shows the presence of uterine fibroids then appropriate tx should be planned based on size, number and location of the fibroids.
- Saline infusion sonography and MRI should NOT be used as a 1st-line diagnostic tool.
- Dilatation and curettage alone should NOT be used as a diagnostic tool.

Education and information provision

A woman with HMB referred to specialist care should be given info before her OPC appt. Women should be made aware of the impact on fertility that any planned surgery or uterine artery embolisation (UAE) may have, and if a potential tx (hysterectomy or ablation) involves the loss of fertility then opportunities for discussion should be made available. Give the following info on potentially unwanted outcomes.

Pharmaceutical treatment

- If hx and ix's indicate that pharmaceutical tx is appropriate and either hormonal or non-hormonal txs are acceptable, txs should be considered in the following order:
- levonorgestrel-releasing intrauterine system (LNG-IUS) provided long-term (at least 12-mos)
- tranexamic acid or NSAIDs or COCs
- norethisterone 15 mg od from days 5 to 26 of menses, or injected long-acting progestogens.
- If hormonal tx's are not acceptable to the patient, then use either tranexamic acid or NSAIDs.
- When HMB coexists with dysmenorrhoea, NSAIDs should be preferred to tranexamic acid.
- Ongoing use of NSAIDs and/or tranexamic acid is recommended for as long as it is beneficial.
- Use of NSAIDs and/or tranexamic acid should be stopped if it does not improve sxs within 3 cycles.
- Use of a gonadotrophin-releasing hormone analogue could be considered prior to surgery or when all other tx options for uterine fibroids, including surgery or uterine artery embolisation, are

contraindicated. If this tx is to   be used for > 6 months or if adverse effects are experienced then HRT 'add-back' therapy is recommended.

- Danazol and Etamsylate should NOT be used routinely for the treatment of HMB.
- Oral progestogens given during the luteal phase only should NOT be used for the treatment of HMB.

Non-hysterectomy surgery

In women with HMB alone, with uterus < than a 10-week pregnancy, endometrial ablation should be considered preferable to hysterectomy. Women must be advised to avoid subsequent pregnancy and on the need to use effective contraception, if required, after endometrial ablation.
Endometrial ablation should be considered in women with HMB who have a normal uterus and also those with small uterine fibroids (< 3 cm in diameter).

UAE, myomectomy or hysterectomy should be considered in cases of HMB where **large fibroids** (> 3 cm in diameter). Women should be informed that UAE or myomectomy may potentially allow them to retain their fertility. Pretreatment before hysterectomy and myomectomy with a GnRH-analogue for 3/12 to 4/12 should be considered where uterine fibroids are causing an enlarged or distorted uterus.

If a woman is being treated with gonadotrophin-releasing hormone analogue and UAE is then planned, the gonadotrophin-releasing hormone analogue should be stopped as soon as UAE has been scheduled.

Hysterectomy - Taking into account the need for individual assessment, the route of hysterectomy should be considered in the following order: 1st line vaginal; 2nd line abdominal.

The Mental Health Act 2007 (the 2007 Act)

Amends the Mental Health Act 1983 (the 1983 Act), the Mental Capacity Act 2005 (MCA) and Domestic Violence, Crime and Victims Act 2004.

Amendments to the Mental Health Act 1983 to ensure that persons with serious mental disorders which threaten their health or safety or the safety of the public can be treated irrespective of their consent where it is necessary to prevent them from harming themselves or others.

- **Definition of mental disorder**: 1983 Act changed so that a single definition applies throughout the Act, and abolishes references to categories of disorder.
- **Criteria for detention**: introduces a new appropriate medical treatment test which applies to all the longer-term powers of detention. As a result, it is not possible for patients to be compulsorily detained, or their detention continued, unless appropriate medical treatment and all other circumstances of the case is available to that patient. These criteria abolished the treatability test.
- **Professional roles**: broadens the group of practitioners who can take on the functions currently performed by the approved social worker (ASW) and RMO.
- **Nearest relative**: gives patients the right to make an application to the county court to displace their nearest relative and enables county courts to displace a nearest relative who it thinks is not suitable to act as such.
- **Nearest relative:** amended to include civil partners amongst the list of relatives.
- **Supervised Community Treatment (SCT):** introduces SCT for patients following a period of detention in hospital. SCT will allow certain patients with a mental disorder to be discharged from detention subject to the possibility of recall to hospital if necessary. This is particularly intended to help avoid situations in which some patients leave hospital and do not continue with their treatment, with the result that their health deteriorates and they require detention again – the revolving door.
- **Electro-convulsive therapy:** introduces new safeguards
- **Tribunal:** reduces the periods after which hospital managers must refer certain patients' cases to the Tribunal if they do not apply themselves.

Introduces an order-making power to make further reductions in due course.

- **Independent mental health advocacy:** places a duty on the appropriate national authority to make arrangements for help to be provided by independent mental health advocates.
- **Age-appropriate services:** will require hospital managers to ensure those aged <18 admitted for mental disorder are in an environment suitable for their age (subject to needs).

2007 Act Amendments to the Mental Capacity Act 2005

The main change is to provide for procedures to authorise the deprivation of liberty of a person in a hospital or care home who lacks capacity to consent to being there. The MCA principles of supporting a person to make a decision when possible, and acting at all times in the person's best interests and in the least restrictive manner, will apply to all decision-making in operating the procedures

MENTAL HEALTH ACT 1983

Section 2   admit for **ASSESSMENT**, compulsory to hospital, up to 28 days, not renewable, requires 2 doctors (one approved under section 12 and one GP), application for admission by social worker or nearest relative (spouse – offspring – parent)

Section 3   admit for **TREATMENT**, compulsory to hospital, up to 6/12, has diagnosis, requires 2 doctors (one section 12 and one GP), application for admission by SW or nearest relative.

Section 4   **EMERGENCY** admission, compulsory to hospital, 72 h, not renewable, seen pt in last 24 hrs, requires 1 doctor if 'urgently necessity' and SW or nearest relative, change to section 2 later

Section 5   **DETAIN IN-PATIENT** by one doctor in emergency only or by nurse in hospital, 72 h. If the doctor is not a psychiatrist, he must act in person to contact a psychiatrist; section 2 or 3 can be considered.

Section 7   appropriate for GUARDIANSHIP (reception into)

Section 12   used to approve doctors recognised as having special expertise in mental health

Section 115   right of entry into premises by SW

Section 135   police right of entry into premises to remove patient to place of safety under a magistrate's warrant

Section 136   police right to remove patient from public place to place of safety (prison cell or hospital) for 72 hrs to allow medical exam.

MULTIPLE SCLEROSIS

WebMD: 28 April 2009: High doses of Vitamin D cut MS Relapses. High dose vitamin D reduces T cell activity (T lymphocytes order attacks on the myelin sheaths that surround and protect the brain cells.). 16% relapsed on 14000 IU vitamin D daily x 1 y vs. 40% on 1000 IU daily. J Burton Neurologist at Univ of Toronto.

NICE Nov 2003/ (add Tysabri)
Three Types
- **Relapsing/ remitting MS** - periods of good health or remission followed by sudden sxs or relapses (80% at onset). On the basis of hx of sxs, is this the 1st attack of presumed demyelination? Yes → does the exam reveal evidence of > a single lesion in the CNS? Yes await 2nd clinical attack or do MRI brain after 3months. No - check for positive CSF (oligoclonal IgG bands in CSF and not serum or elevated IgG index), then go to brain MRI scan - shows ≥ 2 lesions = MS
- **2° progressive MS** - follows on from relapsing/ remitting. Gradually more or worsening sxs with fewer remissions (50% with relapsing/ remitting MS dvlp this during the 1st 10 yrs of illness)
- **1° progressive MS** - from the start sxs gradually dvlp and worsen with time (10-15% at onset). No attacks but progress from onset with 1 objective lesion → Perform LP (positive IgG)→ Brain MRI shows dissemination in space or time (Gd-enhancing lesion at a site different from attack 3/12 ago).

General Principles
- Communication (straightforward, check understanding). Emotional support. Encouraging autonomy/ self-mx. Support to family and carers - inform about social services carer assessment.
- Provide info on activities that promote health maintenance and prevent complications, changes in health that may require further action, condition and its mx.
- Specialist neurological service for: dx of MS initially and of subsequent sxs as necessary, provision of disease-modifying drugs and enacting th risk-sharing scheme for interferon beta and glatiramer acetate.
- Specialist neurological rehab service - assessment, integrated programme of rehab, monitor change, give advice to other services; comprised of   docs, nurses, physio, OT, speech and language

herapists, clinical psychologists, social workers. Teamwork.
- Ready access to dietetics, liaison psychiatry, continence advisory and mx services, pain mx services, chiropody, podiatry, ophthalmology
- People with MS should be able to identify and contact named person in health care area who is responsible for all NHS services for local people with MS and a named person with clinical expertise

Individuals who are severely impaired should have their needs met as follows: Additional support in the home. Respite care in the home. Respite care in another age-appropriate setting. Moving into a residential or nursing home

Diagnosis
- Inform of the potential dx of MS as soon as is likely, before undertaking further ixs to confirm or refute dx.
- Find out what and how much info the pt wants to receive.
- Discuss the nature and purposes of all ixs, the likely outcomes and their implications.
- Confirmed diagnosis of MS should be told by a doctor with specialist knowledge about MS (consultant or experienced SpR) and offer at least one more appointment with the doctor who gave the original dx.
- Put in touch with specialist MS skilled nurse or support worker with counselling experience.
- Offer written info about local and national disease-specific support orgs, including local rehabilitation services.
- Offer info about the disease, preferably in an info pack.
- Within 6/12 of the diagnosis, offer the opportunity to participate in an educational programme to cover all aspects of MS
- When a patient presents with a 1st episode of neurological sxs or signs suggestive of demyelination, consider MS.
- When an individual presents with a 2nd or subsequent set of neurological sxs which are potentially attributable to inflammatory or demyelinating
  lesions in the CNS, refer to an expert for ix.
- The dx should be made clinically by a doctor with specialist neurology experience, on the basis of evidence of CNS lesions scattered in space and time, primarily on the basis of hx and exam.

- When the dx remains in doubt, further investigation should exclude alternate dx or find evidence to support dx (dissemination in space or time - MRI, dissemination in space - visual evoked potential studies, CSF analysis - for presence of oligloconal bands and compare with serum samples).
- The diagnosis of MS is clinical and an MRI scan should not be used in isolation to make the diagnosis.

Diagnosis and treatment of an acute episode

- Sudden (within 12-48h) ↑ in neurological sxs or disability or develops new neurological symptoms, a formal assessment should be made to determine the diagnosis.
- Diagnosis of optic neuritis - acute, sometimes painful, reduction or loss of vision in 1 eye. Common presenting sign. Acute decline in VA (+/- pain) should   be seen by ophthalmology for diagnosis. If the diagnosis is confirmed, the ophthalmologist should discuss the potential diagnosis and offer a referral to neurologist.
- Tx: IV methylprednisolone 500 mg-1g daily x 3-5/7 OR high-dose oral methylprednisolone 500 mg - 2g daily for 3-5/7. Discuss risks/ benefits of steroids.
- Dx of transverse myelitis - acute episode of weakness or paralysis of both legs, with sensory loss and loss of control of bowels and bladder is an emergency. Urgent investigation for person presenting with sxs and signs of acute spinal cord dysfunction to exclude surgically treatable compressive lesion.

Interventions affecting disease progression

- Advise Linoleic acid 17-23/g /day may ↓ progress of disability (sunflower, corn, soya, safflower oils)
- Criteria by Assoc of British Neurologists and agreed by DOH to determine eligibility for tx with interferon beta and glatiramer acetate for people with MS within the 'risk-sharing scheme':
- People with relapsing/ remitting MS should be offered interferon beta provided that the following 4 conditions are met: can walk 100m without assistance, had at least 2 clinically significant relapses in the past 2 yrs, are ≥ 18, do not have C/Is.
- People with relapsing/ remitting MS should be offered glatiramer acetate provided that the following 4 conditions are met: can walk 100m

without assistance, at least 2 clinically significant relapses in the past 2 yrs, ≥ 18, or no C/Is.

- People with secondary progressive MS should be offered interferon beta provided that they: can walk 10m without assistance, at least 2 clinically significant relapses in the past 2 yrs, minimal ↑ in disability due to gradual progression over the 2 yrs, ≥ 18, or no C/Is.

- People with MS offered tx with interferon beta OR people with relapsing-remitting MS offered tx with glatiramer acetate should have the following stopping criteria: intolerant S/Es, becoming or trying to become pregnant, occurrence of 2 disabling relapses within a 12-month period, secondary progression with an observable ↑ in disability over a 6-month period, loss of ability to walk +/-assistance for > 6 months.

- Treatments in specific circumstances by expert with close monitoring for adverse events: azathioprine, mitoxantrone, IV immunoglobulin, plasma exchange, intermittent (4-monthly) short (1-9 days) courses of high-dose methylprednisolone

- The following txs should NOT be used due to lack of evidence: cyclophosphamide, anti-viral - aciclovir, tuberculin, cladribine, long-term steroids, hyperbaric O_2, linomide, whole-body irradiation, myelin basic protein

Altering the risk of relapses

- Immunise against influenza. Advise those who wish to become pregnant, that the risk of relapse ↓ during pregnancy and ↑ transiently postpartum. Encourage to have any op they need.

Rehab and maintenance

- New limitation - is it due to an unrelated disease, incidental infection, MS relapse or grad progression?

- Specialist vocational rehab service (disability employment advisers, Access to Work Scheme) entitled to adjustments at work under the Disability Discrimination Act

- Specialist neuro rehab (driving adaptive equip, wheelchairs, physio)

- ADL -personal (wash, dress, eat, toilet), domestic (cook, wash, iron, clean, bills), community (shop, public transport, avoid traffic), caring for children. Share with social services. Paid carers.

Management of specific impairments

- Fatigue - aerobic exercise, energy conserving technique, rx amantadine 200 mg od small evidence
- Bladder dysfunction - post-micturition residual bladder volume (U/S bladder), UTI, intermittent self-cath, rx anticholinergics (oxybutynin or tolterodine) for urge incontinence. Check for ↑ in post-void residual volume. Nocturia rx desmopressin 100-400 ug orally or 10-40 ug intranasally nocte or for travel; must not use > once in 24h period. Special continence service for assessment, pads, convene drain (men), self-cath, l-t urethral cath, IV botulinum toxin (trained doctor).
- UTI - cranberry juice, prophylactic antibiotics, continence specialist assess for residual urine if 3 UTIs in 1yr
- Bowel - oral laxatives, suppositories, enemas
- Spasticity-neuro-physiotherapists, phys techniques to avoid contractures.   Rx 1st-line baclofen or gabapentin. 2nd line - tizanidine, diazepam, clonazepam or dantrolene. Splints, serial casting, intrathecal baclofen.
- Vision - ↓ read newspaper or watch tv due to poor control of eye movements, assess for low-vision equipment and adaptive technology, refer to specialist social services team, register partially sighted
- Neuropathic pain - carbamazepine, gabapentin or amitriptyline, refer
- Musculoskeletal pain - TENS or antidepressant
- Cognitive loss - formal cognitive neuro-psychological assessment with specialist clinical psychologist
- Emotionalism - cry or laugh with min provocation. Rx TCA or SSRI
- Depression - Do you feel depressed? Liaison psychiatrist if severe. Rx antidepressant or CBT
- Anxiety - Rx antidepressant or benzo
- Swallowing - PEG, swallowing technique using videofluoroscopy, short-term NG tube, chest physio
- Speech - dysarthria - speech and language therapist
- Sex- sildenafil 25-100 mg men
- Pressure ulcers - assess if using wheelchair
- Complementary therapies- some evidence - reflexology and massage, fish oils, magnetic field therapy, neural therapy, massage plus body work, t'ai chi, multi-modal therapies.

BMJ Clinical Evidence Neck Pain A Binder M.D.
What are the effects of txs for uncomplicated neck pain without severe neurologic deficit?

Likely to be beneficial
- Manual Treatments (Mobilization and Manipulation). Systematic Reviews and RCTs found limited evidence that manipulation or mobilization improved symptoms vs. other or no treatment in patients with neck pain.
- Physical Therapies (Active Physio, Exercise, Pulsed Electromagnetic Field Therapy). Systematic reviews and RCTs found that active physio ↓ pain vs. passive treatment, and that exercise programs ↓ pain vs. mx that does not include exercise. 1 RCT provided limited evidence that pulsed electromagnetic field treatment ↓ pain vs. sham treatment.

Unknown effectiveness
- Drug Therapies (Analgesics, NSAIDs, Antidepressants, or Muscle Relaxants. Insufficient evidence on the effects of drug txs for neck pain, although they are widely used. Several drugs used to treat neck pain are associated with well-documented adverse effects.
- Multidisciplinary (Multimodal) Tx. RCTs provided insufficient evidence to compare effects of multimodal txs vs. other tx in patients with uncomplicated pain.
- Physical Txs (Heat or Cold, Traction, Biofeedback, Spray and Stretch, Acupuncture, Laser). Systematic reviews found insufficient evidence about the effects of these physical txs.
- **Soft Collars and Special Pillows**. No RCTs of sufficient quality on the effects of soft collars/ pillows.

Unlikely to be beneficial
- Pt Ed. 3 RCTs found no significant difference between pt ed (advice or group instruction) with or without analgesics vs. no tx, stress mx, placebo, or usual care.

What are the effects of txs for acute whiplash injury?
Likely to be beneficial
- Early Mobilization. Systematic reviews and subsequent RCTs provided limited evidence that early mobilization ↓ pain vs. immobilization or rest plus a collar.
- Early Return to Normal Activity. Systematic reviews and subsequent RCTs provided limited evidence that advice to "act as usual" + NSAIDs improved mild symptoms vs. immobilization + 14 days of sick leave.

- Electrotherapy. 1 small RCT provided limited evidence that electromagnetic field tx↓ pain after 4/52 but not after 3/12 vs. sham tx.
- Multimodal Treatment. 1 RCT found that multimodal treatment ↓ pain at the end of treatment and after 6/12 vs. physical treatment.

Unknown effectiveness

- Drug Txs. No RCTs of drug txs in acute whiplash injury.
- Home Exercise Programs. 1 RCT found no significant difference between different home exercise programs in pain or disability.

What are the effects of treatments for chronic whiplash injury?
Likely to be beneficial

- Percutaneous Radiofrequency Neurotomy. 1 RCT provided limited evidence ↓ pain vs. sham treatment after 27/52.

Unknown effectiveness

- Multimodal Therapy (Physiotherapy + CBT). 1 RCT found no significant difference between multimodal therapy (physio + CBT) in disability, pain, or range of movement at the end of treatment or at 3/12.
- Physiotherapy. 1 RCT found no significant difference between physio alone and multimodal therapy (physio + CBT) in disability, pain, or range of movement at the end of treatment or at 3/12.

What are the effects of treatments for neck pain with radiculopathy?
Unknown effectiveness

- Drug Thrapies (Epidural Steroid Injections, Analgesics, NSAIDs, or Muscle Relaxants). We found no RCTs on the effects of epidural steroid injections, analgesics, NSAIDs, or muscle relaxants.
- Surgery vs. Conservative Treatment. 1 RCT found no significant difference in symptoms after 1 year.

BMJ Clinical Evidence: Nocturnal Enuresis
What are the effects of interventions for the relief of sxs?
Beneficial

Enuresis Alarm. 1 systematic review showed that an alarm ↓ wet nights during tx and ↑ tx success rates at 10 to 20 wks vs. no tx. 1 review showed no significant difference between alarms and dry-bed training in tx success rates at 4 to 20 weeks. The review showed that an alarm + dry-bed training ↑ treatment success and ↓ relapse at 8 to 20 weeks vs. no treatment. There were no significant differences between an alarm and dry-bed training and an alarm alone in tx success or in relapse rates after discontinuing tx. Another systematic review showed no significant difference between desmopressin and an alarm in children who achieved initial success. However, it showed limited evidence that desmopressin ↓ wet nights vs. an alarm during the 1ˢᵗ week of treatment. A subsequent RCT showed that desmopressin significantly ↓ wet nights during 12 weeks of treatment vs. an alarm. 1 systematic review showed that desmopressin + alarm ↓ wet nights per week during treatment vs. an alarm alone or an alarm + placebo. However, there was no significant difference between an alarm + desmopressin and an alarm alone in treatment success rates (defined as 14 consecutive dry nights) after 6 to 24 weeks or in relapse rates after tx discontinuation. RCTs showed that enuresis alarms ↓ treatment failure and relapse rates after treatment discontinuation at 8 to 14 weeks vs. tricycles. 1 RCT showed no evidence that an enuresis alarm was more effective after 5 to 6 weeks when combined with tricyclics.

Dry-Bed Training Plus Enuresis Alarm. ↑ treatment success rates and ↓ relapse rates.

Desmopressin (while treatment continues). 1 systematic review showed that desmopressin ↓wet nights and ↑ initial treatment success vs. placebo. The review showed insufficient evidence comparing intranasal with oral desmopressin or desmopressin with tricyclics. There was limited evidence that higher doses of desmopressin ↓ wet nights during treatment vs. lower doses. 1 systematic review showed insufficient evidence comparing desmopressin with tricyclics. The review showed no significant difference between desmopressin and an alarm in children achieving initial success. It showed limited evidence that desmopressin ↓ wet nights vs. an alarm during the 1ˢᵗ week of treatment.
A subsequent RCT showed that desmopressin ↓ wet nights during 12 weeks of treatment vs. an alarm. 1 RCT showed no significant difference between laser acupuncture and intranasal desmopressin in the number of wet nights after 3/12. No RCTs compared effects of adding desmopressin to anticholinergics oxybutynin, tolterodine, or hyoscyamine.

Trade-off between benefits and harms

Tricyclic Drugs (Imipramine, Desipramine). 1 systematic review showed that the use of tricyclics (imipramine, desipramine) ↑ treatment success rates vs. placebo, although this benefit did not continue after tx discontinuation. Tricyclics are associated with adverse effects (anorexia, anxiety, constipation, depression, diarrhoea, dizziness, drowsiness, dry mouth, headache, irritability, lethargy, sleep disturbance, upset stomach, vomiting) vs. placebo.
RCTs showed that an enuresis alarm ↓ treatment failure and relapse after tx discontinuation at 8 to 14 weeks vs. tricyclics. 1 RCT showed no evidence that

an alarm was more effective after 5 to 6 weeks when combined with tricyclics. 1 systematic review showed insufficient evidence comparing desmopressin with tricyclic drugs. One small RCT showed that a combination of oxybutynin and imipramine ↑ treatment success vs. placebo.

Unlikely to be beneficial

Desmopressin (After Tx Discontinuation). Small RCTs showed limited evidence that there was no significant difference between desmopressin and placebo in tx success rates after tx discontinuation. 2 RCTs showed that desmopressin was less effective than alarms when tx ended.

Standard Home Alarm Clock. 1 RCT showed more children using standard home alarm clocks achieved 14 consecutive dry nights during 4/12 of tx vs. those who were awakened every 3h; however, it showed no significant difference in relapse rates 3/12 after tx discontinuation.

Dry-Bed Training. 1 systematic review showed insufficient evidence to determine whether dry-bed training alone for 8 to 24 wks was superior to nil. The review showed that an alarm ↑ dry nights at 4 to 20 wks vs. dry-bed training alone.

Desmopressin + Enuresis Alarm. 1 systematic review showed desmopressin + enuresis alarm ↓ wet nights per wk during initial tx vs. alarm alone or the alarm + placebo. There was no significant difference between the alarm and desmopressin and the alarm alone in tx success rates at 6 to 24 wks or in relapse rates after tx discontinuation.

Anticholinergic Drugs (Oxybutynin, Tolterodine, Hyoscyamine). 1 small RCT provided insufficient evidence to determine the effectiveness of the anticholinergics for the tx of nocturnal enuresis. We found no RCTs comparing oxybutynin, tolterodine, and hyoscyamine with desmopressin, tricyclic drugs, alarms, or dry-bed training. 1 small RCT showed that combining oxybutynin and imipramine ↑ tx success rates vs. placebo.

Laser Acupuncture (as Effective as Desmopressin in 1 RCT). 1 RCT showed no significant difference between laser acupuncture and intranasal desmopressin in the no. of wet nights after 3/12.

Page | 198

NOTIFIABLE DISEASES
under the Public Health Act of 1984 and Public Health Regs 1988
Due to strong suspicions of a linkage between vCJD and BSE, the British government made BSE a notifiable disease in 1988. **CJD** (transmissible spongiform encephalopathies) became a notifiable disease in 1999.

anthrax
cholera
diphtheria
dysentery (amoebic or bacillary)
encephalitis (includes TSE)
food poisoning (suspected or proven)
lassa fever
leprosy
leptospirosis
malaria
Marburg disease

measles
meninigitis
mumps
ophthalmia neonatorum
paratyphoid A or B
plague
polio
rabies
relapsing fever
rubella

scarlet fever
smallpox
tetanus
Tb
typhoid fever
typhus fever
viral haemorrhagic fever
viral hepatitis
whooping cough
yellow fever

OBSTETRICS

Abortion TOP, spontaneous or induced, prior to 24 wks' gest

Incomplete abortion PVB in early pregnancy associated with passage of products of conception + open cerv os

Inevitable abortion PVB in early pregnancy associated with open os and NO passage of products of conception

Missed abortion small for dates uterus with closed cervical os; fetal death

Septic abortion any type of abortion that becomes infected

Threatened abortion PV spotting in early pregnancy with a closed cervical os and NO passage of POC

Placenta abruptio antepartum PVB after 24 wks' gestation associated with lower abdo pain due to separation of the placenta before delivery of the infant

Placenta praevia antepartum painless recurrent PVB after 24 weeks' gestation due to a low lying placenta; vaginal exam must not be performed if suspect praevia.

Postpartum haemorrhage 1^0 – blood loss $\geq$ 500 ml within 24h of delivery
2^0 – blood loss $\geq$ 500 ml after the 1st 24 hours; usually due to infection or RPC

OCCUPATIONAL HEALTH FOR HEALTHCARE WORKERS
DOH Guidance

HEPATITIS A Maintenance workers involved in procedures likely to involve contact with raw sewerage, drain cleaning /unblocking. Discuss exposure risks with Occupational Health who will offer Hepatitis A vaccination if necessary.

HEPATITIS B DOH guidance on Hep B imms must be followed and incorporated into a policy for staff immunization. Vaccination will be offered to all staff at employment whose occupation may involve contact with blood and body fluids, unless they can show clear evidence of immunity. In particular, staff undertaking Exposure Prone Procedures (EPP's) will be required to provide evidence of immunity. The primary course consists of 3 injections given over 6 mos. A blood sample will be taken on completion of the course to ensure adequate response to the vaccine. In the event of a sharps injury or significant exposure to blood or blood-contaminated body fluids e.g. conjunctival splash, the degree of risk will be assessed and appropriate prophylaxis may be provided. It is essential that all accidental exposures are reported ASAP, as in certain circumstances it is necessary to give Hep B immunoglobulin and/or booster doses of vaccine, even though the individual has been immunised, to minimise the risk of acquiring infection.

HEPATITIS C Follow DOH guidance on Hep C. Currently, no preventative vaccine is available for HCV.

POLIOMYELITIS A basic course of oral polio vaccine, consisting of 3 doses will be offered to all staff, who are uncertain of their previous vaccination hx, when they commence employment. A booster dose will then be offered at appropriate intervals thereafter. Staff receiving booster doses will be reminded that they may be infectious to family contacts and patients who are currently unprotected. The Occupational Health Dept staff will give relevant advice. Booster doses are recommended when leaving school or starting employment and every 10 years thereafter.

RUBELLA (GERMAN MEASLES) A hx of rubella is discounted unless confirmed by a blood test. Staff who may contract rubella could unknowingly pass it on to susceptible individuals i.e. immune-suppressed patients and pregnant women. Such staff should be immunised. Rubella vaccine is recommended for all sero-negative female staff of childbearing age and all sero-negative staff, male or female, who are in contact with patients. Where necessary staff will be screened on employment and the rubella vaccine will be offered to those who are non immune. Rubella vaccine is contra-indicated in pregnancy and pregnancy must be avoided for 1 month after vaccination.

TETANUS Active immunisation against tetanus is effective in the prevention of disease and it will be offered to all staff who are assessed as not being protected. A basic course consists of 3 injections and 2 boosters, which is usually completed at the end of an individual's schooling. It is very important that staff involved in gardening and maintenance are immunised as they are most at risk, however all staff will be encouraged to be protected against tetanus. Staff who have previously received 5 doses will not be offered a booster dose other than following an injury at risk of tetanus.

TUBERCULOSIS The major source of Tb is from individuals who have active pulmonary disease with infected sputum. It is uncommon for health care staff who are in good health, to acquire Tb from patients. All staff including students in regular contact with patients/clients and those who handle material which may contain tubercle bacilli (sputum specimens) are at risk.

The Occupational Health Dept will screen all staff prior to or soon after taking up employment. The nurse will enquire about past BCG vaccination, previous Heaf / Mantoux testing and any recent cxrs. Re-screening will be carried out via Heaf test and cxr if deemed appropriate.

Under new DOH guidance, new staff who refuse Heaf testing, BCG vaccination or cxr (if deemed necessary) may find themselves at risk of contravening trust policy and may be unable to take up a contract of employment. Staff may also be required to co-operate in any contact tracing / screening programme if exposure to an infected pt is identified by the Tb specialist team.

INFLUENZA It is now recognized that this is a cost-effective way of ensuring that key clinical staff, remain fit and well during the influenza season. Decisions on staff immunizations will be made annually by the Trust following the guidance of the CCDC and in line with nat'l guidance. The Occu Health Dept will advise Trust and staff on the appropriate routes for obtaining the vaccine.

WELL'S CRITERIA FOR DVT (Possible score -2 to 8)

Scarvelis D, Wells P (2006). "Diagnosis and treatment of deep-vein thrombosis". *CMAJ* 175 (9): 1087–92.

1.	Active cancer (treatment within last 6/12 or palliative)	+1
2.	Calf swelling > 3 cm compared to other calf (measure 10 cm below tibial tuberosity)	+1
3.	Collateral superficial veins (non-varicose)	+1
4.	Pitting oedema (greater in the symptomatic leg)	+1
5.	Swelling of entire leg	+1
6.	Localized pain along distribution of deep venous system	+1
7.	Paralysis, paresis, or recent cast immobilization of lower extrem	+1
8.	Recently bedridden > 3 days, or major op requiring regional or GA in past 4 weeks	+1
9.	Previous documented DVT	+1
10.	Alternative diagnosis at least as likely	-2

Interpretation: Score of ≥ 2 DVT is likely. Consider imaging the leg veins. Score of < 2 DVT is unlikely. Consider d-dimer test to further rule out DVT.

WELL'S CRITERIA FOR PULMONARY EMBOLISM

Neff MJ (2003). "ACEP releases clinical policy on evaluation and management of pulmonary embolism". *American family physician* 68 (4): 759–60.

• clinically suspected DVT	+3
• alternative diagnosis is less likely than PE	+3
• tachycardia	+1.5
• immobilization/surgery in previous four weeks	+1.5
• history of DVT or PE	+1.5
• hemoptysis	+1.5
• malignancy (treatment for within 6 months, palliative)	+1

Traditional interpretation
Score > 6 High (probability 59% based on pooled data)
Score 2 to 6 Moderate (probability 29% based on pooled data)
Score < 2 Low (probability 15% based on pooled data)

Alternate interpretation
Score > 4 PE likely. Consider diagnostic imaging.
Score ≤ 4 PE unlikely. Consider D-dimer to rule out PE.

ORTHOPAEDICS/ SPORTS INJURIES

KNEE
Prepatellar bursitis – housemaid's knee from repeated kneeling
Patella dislocation – patella apprehension test; reduced by slowly extending knee while applying pressure to lateral aspect of patella
Patellar tendonitis – jumper's knee; NSAIDs, stretching quads and hamstrings
Osgood-Schlatter – injury in tibial apophysis by overly tight patellar tendon and ligament; growth spurt F – 8-13, M 10-15; ice pack post exercise + stretch quads and hamstrings
MCL injury – lat blow to knee (slide tackle in football) MCL attached to medial meniscus (also injured); MCL more common than **LCL**
ACL injury – turning motion of body or ext force over fixed lower leg; elicit pain by pressure on either side of patellar tendon
PCL injury – from hyperextension
Iliotibial band injury – band runs from tensor fascia lata muscle to proximal lateral tibia; IT synd – band compresses lat femoral epicondyle. NSAIDs + stretch band and hip flexors
Medial meniscal tear - + **McMurray test** 'hear click' – knee is flexed and put in valgus stress. Other hand externally rotates leg while extending knee.
Lateral meniscal tear – 'click' when medial knee is stabilized and leg is internally rotated and knee extended.

ANKLE/ FOOT
Achilles tendon rupture - 30-50 yo forceful push-off of foot while extending knee. With the pt prone and calf squeezed, there is absence of passive plantar flexion of foot - Thompson test + for Achilles tendon rupture.
Ankle sprain – inversion> eversion and occurs when foot is plantar flexed and inverted; may damage lateral ligaments.
Plantar fasciitis – RF: pes cavus and runners. Heel pain worse initially in morning and when walking barefoot. Prefers to walk on toes. Tender on anteromedial aspect of heel – plantar fasciitis. Pain is worse on passive dorsiflexion of toes. NSAIDs, stretch Achilles tendon and toes, figure 8 arch sling support.
Morton's neuroma - c/o burning and shooting pain in 3rd intermetatarsal area, relieved by massage.
Navicular stress fracture - track and field runner c/o midfoot pain. Tender over dorsum at midfoot (N spot).

DISC PROLAPSE/ NERVE DISTRIBUTION/ (MEDIAN+RADIAL NERVE INJURIES)

L5/S1 Disc Prolapse Pain along post thigh with radiation to the heel; Weakness on plantar flexion (may be absent); Sensory loss in the lateral foot; Absent ankle jerk reflex.

L4/L5 Disc Prolapse Pain along the post or posterolateral thigh with radiation to dorsum of foot; Weakness of dorsiflexion of the big toe and foot; Paraesthesia and numbness of dorsum foot + big toe; No reflex changes.

L3/L4 Disc Prolapse Pain in front of thigh; Wasting of quadriceps muscles may be present; ↓ sensation on the front of the thigh and medial lower leg; ↓ knee jerk reflex.

Common peroneal nerve (L4-S2) also known as lat popliteal nerve. Lesions lead to equinovarus. Inability to dorsiflex feet and toes. Sensory loss dorsum foot.

Tibial nerve (S1-3) Loss causes calcaneovalgus. Inability to stand on tiptoe or invert foot. Sensory loss sole.

OTTAWA ankle and foot rules
(BMJ 2003;326 (7386):417; BMJ 1995, Bandolier 1998)

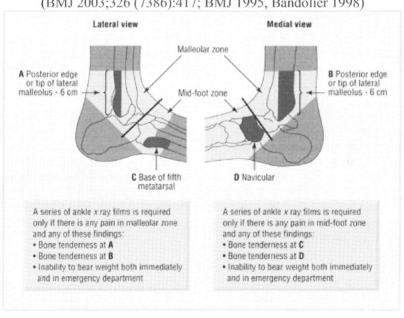

A foot x-ray is required if there is any pain in the midfoot zone and any of these findings: bone tenderness at C, bone tenderness at D, inability to weight bear both immediately and in A+E.

Ottawa knee rules
A knee x-ray is only required for knee injury patients with any of these:
* age ≥ 55
* isolated tenderness of the patella (no bone tenderness of the knee other than the patella)
* tenderness at the head of the fibula
* inability to flex to 90°
* inability to weight bear both immediately and in the casualty department (4 steps - unable to transfer weight twice onto each lower limb regardless of limping).

OSTEOPOROSIS
↓ bone mass+ micro architectural bone deterioration. **WHO definition**: hip or spine density **T score<-2.5 (-2.5SD)**
* Susceptible to fractures in hips (> 75 yo F), vertebrae (associated with steroid use so give osteoporosis protection) and wrists (Colles in > 65 yo F).
* Post-menopausal F loses 2% per yr in bone mass for the 1st 8 years and then 1% per yr. By age 80, she will have lost 30-40% of her bone mass.

RFs: previous fracture from minor fall ; FH of osteoporosis (2x risk); a natural or early surg menopause < age 45; pre-menopausal amenorrhoea > 6 mos not due to pregnancy; alcoholism; liver disease; smoking; prolonged bed rest; malabsorption; thyroid disease; male hypogonadism; rheum arthritis (on systemic steroids) steroid use (> 7.5 mg prednisolone od for > 3/12, i.e. RA, PMR (50 mg prednisolone od and tapered over 18 months), asthma, COPD (short bursts of steroids do not ↑ risk), ulcerative colitis).
Ix: TFTs, vitamin B level and ESR. Fat is protective.

Prevention:

* Avoid exercise-related amenorrhoea. Limit alcohol intake. Stop smoking. Regular weight-bearing exercise.
* Give all > 75 yo F, calcichew D3 Forte q day or reg calcichew 2/ day.

- Maintain recommended daily intake of 700 mg calcium and 400 IU vitamin D (90% from the sun). Post-menopausal F requires 1.5 g of calcium.

Hip fracture

- 20% of patients with hip fractures die w/n a yr after have this major op (NSF 2001).
- 50% lose their independence and need to rely on a stick or nursing home placement (NSF 2001)
- 60% are unable to walk independently at 12 metres (NICE 2000).
- Average age of pt at risk is 80 years.

Vertebral fractures

- Often painless. 1st sign may be shrinkage or loss of inches in height, i.e. dowager's hump.
- Elicit pain by asking pt to sneezer or cough and if the pain radiates around the ribs and waist to the front, highly suspicious of vertebral fracture.

Medication:

- Bisphosphonates (i.e. once weekly 2nd gen bisphosphonate - alendronate (fosomax). Give with calcichew D3 Forte. Protective against both hip and vertebral fractures.) Alternative is risedronate, which ↑ bone density and purported to ↓ incidence of hip fractures by 40%.
- Raloxifene – drug of choice for patients with h/o breast CA, FH or MI. It is a synthetic oestrogen receptor modulator. MORE (multiple outcomes of raloxifene) double-blinded placebo-controlled trial in JAMA, 1999   showed risk reduction of both breast cancer and vertebral fractures.

Refer:

- Kyphosis or loss in height; Osteoporosis +/ - fracture; Osteopoenia on x-ray; Multiple RFs; Prolonged used of steroids > 3 months.

RHEUMATOID ARTHRITIS

- Age of onset – 20-35 yo. Risk: 3 x F > M. Symmetrical arthritis involving MCP, MTP, PIP and wrists. Hallmark sxs: early morning stiffness. Fatigue.

Ixs: FBC; ESR; CRP (very sensitive to inflammation and is superior to ESR for diagnosing PMR)
- Rheumatoid factor (positive in 70%; 30% are normal with a + RF and 30% are false negative); Biochemistry
- X-ray negative in early RA; **anti-CCP** (citric-citrullinated peptide) **antibody** & anti-MCV tests (99.7% specific).

Associated with poor prognosis:
- Elderly; Female; High ESR, CRP, alkaline phosphatase and platelets
- Multiple affected joints at onset; Positive rheumatoid factor, the higher the worse the prognosis
- Systemic involvement (iritis, pleural effusion and vasculitis)

Tx: Early referral to rheumatologist
- Aggressive 1st-line treatment with **methotrexate (7.5 mg)** + salazopyrine + hydroxychloroquine +/- prednisolone. Drug combinations are used to switch the disease off and prevent erosions.
- Newer disease-modifying drugs -infliximab (injections per 2 months); etanerceb £10,000 / yr.

Dx and Mx of PARKINSON'S DISEASE in 1° and 2° Care
(NICE June 2006)

PD-progressive neurodegenerative movement disorder resulting from the death of the DA-containing cells of the substantia nigra. There is no reliable test that can distinguish PD from other conditions. The dx is 1° clinical, based on hx and exam. Classically present with bradykinesia, rigidity and rest tremor. Parkinsonism can also be caused by drugs, multiple cerebral infarction and degenerative conditions, i.e. progressive supra-nuclear palsy (PSP) and multiple system atrophy (MSA)).
PD frequently develop psychiatric problems, i.e. depression and dementia. Autonomic disturbances and pain may later ensue, and progress to cause significant disability and handicap with impaired quality of life.↑ prevalence with age and ↑ prevalence + incidence in men.

Patient-centred care-PD patients should have the opportunity to make informed decisions about their care and tx.

Refer patients with suspected PD quickly and untreated to a specialist with expertise in the diff dx of this condition. The Guideline Development Group considered that **suspected mild PD should be seen within 6/52** but new referrals in **later disease with more complex problems require an appointment within 2/52.**

The GDG advise that patients should be seen at regular intervals of 6–12 months to review their dx and reconsider if atypical clinical features develop. Acute levodopa and apomorphine challenge tests should not be used in the diff dx of parkinsonian syndromes. **Communication with PD and their carer** should be aimed towards empowering them to participate in the judgements and choices about their own care. Discussions should be aimed at achieving a balance between the provision of honest realistic info about the condition and the promotion of a feeling of optimism. Because patients may develop impaired cognitive ability, a communication deficit and/or depression, they should be provided with: both oral and written communication throughout the course of the dis. Give families and carers info about the condition, their entitlements to care assessment and the support services available. Patients should have a comprehensive care plan agreed between the individual, their family and/or carers and specialist and 2° healthcare providers and should be offered an accessible point of contact with specialist services. This could be provided by a PD nurse specialist.

Advise PD patients who drive to **inform the DVLA** + their car insurer of their condition at the time of dx.
Dx Suspect PD in patients presenting with tremor, stiffness, slowness, balance problems and/or gait disorders.
Expert vs. non-expert dx Refer suspected PD quickly and untreated PD to a specialist with expertise in the differential dx of this condition.
Clinical vs. post-mortem dx - should be diagnosed clinically and based on the UK Parkinson's Disease Society.

Brain Bank Criteria
Step 1: bradykinesia and at least 1 of the following: muscular rigidity, rest tremor (4-6 hz), postural instability unrelated to primary visual, cerebellar and vestibular or proprioceptive dysfunction.

Step 2: r/o h/o repeated strokes with stepwise progression, repeated head injury. Antipsychotic or DA-depleting drugs, definite encephalitis +/or oculogyric crisis on no drug tx, > 1 affected relative, sustained remission, negative response to large doses of levodopa (if malabsorption excluded), strictly unilateral features after 3 yrs, other neurological features: supranuclear gaze palsy, cerebellar signs, early severe autonomic involvement, Babinski sign, early severe dementia with language disturbance, memory or praxis, exposure to known neurotoxin, cerebral tumour or communicating hydrocephalus on neuroimaging.

Step 3 Supportive criteria for PD. ≥ 3 required for dx of definite PD: unilateral onset, rest tremor present, progressive d/o, persistent asymmetry affecting the side of onset most, excellent response to levodopa, severe levodopa-induced chorea, levodopa response for > 5 yrs, clinical course < 10 yrs.

Discuss with patients the possibility of tissue donation to a brain bank for diagnostic confirmation and research.

Review dx regularly and re-considered if atypical clinical features develop.

Single photon emission computed tomography (SPECT) ^{123}I-FP-CIT SPECT should be considered for patients with tremor where essential tremor cannot be clinically differentiated from parkinsonism.

Drug tx in early PD 'early disease' refers to PD who have developed functional disability and require symptomatic tx. 'Later disease' refers to PD on **levodopa** who have developed **motor complications**. No single DOC in early PD.

Table 1 Options for initial pharmacotherapy in early PD

Initial therapy for early PD	1st-choice option	Symptom control	Risk of side effects	
			Motor complications	Other adverse events
Levodopa	✓	+++	↑	↑
DAs	✓	++	↓	↑
MAO-B inhibitors	✓	+	↓	↑
Anticholinergics	✗	Lack of evidence	Lack of evidence	Lack of evidence
B-blockers	✗	Lack of evidence	Lack of evidence	Lack of evidence
Amantadine	✗	Lack of evidence	Lack of evidence	Lack of evidence

KEY+++ = Good degree of symptom control
++ = Moderate degree of symptom control
+ = Limited degree of symptom control
↑ = Evidence of increased motor complications/other adverse events
↓ = Evidence of reduced motor complications/other adverse events

It is not possible to identify a universal 1st-choice drug tx for early PD.
Levodopa may be used as a symptomatic tx for early PD. The dose of levodopa should be kept as low as possible to **maintain good function in order to ↓ the development of motor complications**.

Dopamine agonists may be used as a symptomatic tx for early PD. Should be titrated to a clinically efficacious dose. If side effects prevent this, use another agonist or a drug from another class.

If an **ergot-derived dopamine agonist** is used, the patient should have renal function tests, ESR and CXR performed before starting tx, and annually thereafter. In view of the monitoring required with ergot-derived dopamine agonists, a non-ergot-derived agonist should be preferred in most cases.

MAO-B inhibitors may be used as a symptomatic tx for early PD.
B-blockers may be used in the symptomatic tx of selected patients with postural tremor, but should not be 1st DOC.

Amantadine may be used as a tx for early PD but should not be 1st drug of choice.

Anticholinergics may be used as a symptomatic treatment in young patients with early PD and severe tremor, but should not be drugs of 1st choice due to limited efficacy and the propensity to cause neuropsychiatric side-effects.
MR levodopa preps should not be used to delay the onset of motor complications in patients with early PD.

Pharmacological tx in later PD Most will develop motor complications and will eventually require levodopa. Adjuvant drugs to take alongside levodopa have been developed with the aim of reducing these motor complications and improving quality of life. There is no single DOC in the pharmacotherapy of later PD.

Table 2 Options for adjuvant pharmacotherapy in later PD

Adjuvant therapy for later PD	1st - choice option	Symptom control	Risk of side effects	
			Motor complications	**Other adverse events**
Dopamine agonists	✓	++	↓	↑
COMT inhibitors	✓	++	↓	↑
MAO-B inhibitors	✓	++	↓	↑
Amantadine	✗	NS	↓	↑
Apomorphine	✗	+	↓	↑

KEY

+++ = Good degree of symptom control   ++ = Mod degree of symptom control

+   = Limited degree of symptom control

↑ = Evidence of increased motor complications/other adverse events

↓ = Evidence of reduced motor complications/other adverse events

NS = Non-significant result

It is not possible to identify a universal 1st-choice adjuvant drug tx for people with later PD. The choice of adjuvant drug 1st prescribed should take into account: clinical and lifestyle characteristics, pt preference, after the pt has been informed of the short- and long-term benefits and drawbacks of drugs.

Modified-release levodopa preps may be used to ↓ motor complications in later PD, but should not be 1st choice. DA agonists may be used to ↓ motor fluctuations in later PD. If an ergot-derived DA agonist is used, the pt should have a min of renal function tests, ESR and CXR before starting tx, and annually.

A DA agonist should be titrated to a clinically efficacious dose. If side effects prevent this, then another agonist or a drug from another class should be used in its place. In view of the monitoring required with ergot-derived dopamine agonists, a non-ergot-derived agonist should be preferred, in most cases.

MAO-B inhibitors may be used to reduce motor fluctuations in later PD.

Catechol-O-methyl transferase inhibitors may be used to ↓ motor fluctuations in later PD.

In view of problems with ↓ concordance, patients with later PD taking entacapone should be offered a triple combination preparation of levodopa, carbidopa and entacapone.

Tolcapone should only be used after entacapone has failed in later PD due to lack of efficacy or side effects. LFTs are required q 2 wks during the 1st year of tx, and thereafter.

Amantadine may be used to ↓ dyskinesia in later PD.

Intermittent apomorphine injections may be used to ↓ off time in PD with severe motor complications.

Continuous subcutaneous infusions of apomorphine may be used to ↓ off time and dyskinesia in PD with severe motor complications. Initiation is restricted to expert units with facilities for appropriate monitoring.

Further drug administration considerations

Antiparkinsonian meds should NOT be withdrawn abruptly or allowed to fail suddenly due to poor absorption (i.e., gastroenteritis, abdo surgery) to avoid the potential for acute akinesia or neuroleptic malignant syndrome. The practice of withdrawing patients from their antiparkinsonian drugs (so called 'drug holidays') to ↓ motor complications should not be undertaken because of the risk of neuroleptic malignant syndrome. In view of the risks of sudden changes in antiparkinsonian med, PD admitted to hospital or care homes should have their med: given at the appropriate times, which in some cases may mean allowing self-medication, adjusted by, or adjusted only after discussion with, a specialist in the mx of PD.

Doctors should be aware of **DA dysregulation syndrome**, an uncommon d/o in which dopaminergic med misuse is associated with abnormal behaviours (hypersexuality, pathological gambling and stereotypic motor acts). Difficult to manage.

Surgery for PD 'Deep brain stimulation for PD' NICE.

Bilateral subthalamic nucleus (STN) stimulation may be used in PD who: have motor complications that are refractory to best medical tx, are biologically fit with no clinically significant active comorbidity, are levodopa responsive and have no clinically significant depression or dementia.

Bilateral globus pallidus interna (GPi) stimulation may be used in PD who: have motor complications that are refractory to best medical tx, are biologically fit with no clinically significant active comorbidity, are levodopa responsive and have no clinical depression or dementia.

Comparison of STN and GPi stimulation for PD With the current evidence it is not possible to decide if the subthalamic nucleus or globus pallidus interna is the preferred target, effective or safer for deep brain stimulation.

Thalamic deep brain stimulation may be considered as an option in PD who predominantly have severe disabling tremor and where STN stimulation cannot be performed.

Non-motor features of PD

MH problems - Depression - should have a low threshold for diagnosing depression in PD. There are difficulties in diagnosing mild depression in PD because the clinical features of depression overlap with the motor features of PD. The mx of depression in PD should be tailored to their co-existing tx.

Psychotic sxs All PD and psychosis should receive a general medical evaluation and tx for any precipitating condition. Consideration should be given to withdrawing gradually antiparkinsonian med that might have triggered psychosis in PD. Mild psychotic sxs in PD may not need to be actively treated if they are well tolerated by the pt and carer. Typical antipsychotic drugs (i.e. phenothiazines and butyrophenones) should NOT be used in PD because they exacerbate the motor features of the condition.

Atypical antipsychotics may be considered for tx of psychotic sxs in PD, although the evidence base for their efficacy and safety is limited. Clozapine may be used in the tx of psychotic symptoms in PD, but registration with a mandatory monitoring scheme is required. Few specialists have experience with clozapine.

Dementia Although cholinesterase inhibitors have been used successfully in individual people with PD dementia, further research is recommended to identify those patients who will benefit from this tx.

Sleep disturbance Take a full sleep hx. Advise **Good sleep hygiene** with any sleep disturbance and includes: avoidance of caffeine in the evening, establish regular pattern of sleep, comfortable bedding and temp, provision of assistive devices (a bed lever or rails to aid with moving and turning), allow the pt to get more comfortable, restrict daytime siestas, advice about taking regular and appropriate exercise to induce better sleep, a review of all rxs and avoidance of any drugs that may affect sleep or alertness, or may interact with other meds (i.e., selegiline, antihistamines, H_2 antagonists, antipsychotics and sedatives). Identify and manage restless legs syndrome (RLS) and REM sleep behaviour disorder. PD who have sudden onset of sleep should be advised not to drive and to consider any occupational hazards. Attempts should be made to adjust their meds to reduce its occurrence. Modafinil may be considered for daytime hypersomnolence in people with PD.

Modified-release levodopa preparations may be used for nocturnal akinesia in people with PD.

For PD at risk of falling, refer to'Falls: assessment and prevention of falls in older people' NICE.

PD should be treated appropriately for the following autonomic disturbances: urinary dysfunction, weight loss, dysphagia, constipation, erectile dysfunction, orthostatic hypotension, sialorrhoea, excessive sweating.

Specialist nurse interventions

PD should have regular access to: clinical monitoring and meds adjustment, a continuing point of contact for support, including home visits, when appropriate, a reliable source of info about clinical and social matters of concern to PD and their carers which may be provided by a PD nurse specialist.

Physio for gait re-education, improvement of balance and flexibility, enhancement of aerobic capacity, improvement of movement initiation and of functional independence, including mobility and ADL, provision of advice regarding safety in the home environment. Offer the Alexander Technique to benefit PD by helping them to make lifestyle adjustments that affect both the phys nature of the condition and the patient's attitudes to PD.

OT for maintenance of work and family roles, home care and leisure activities, improvement and maintenance of transfers and mobility, improvement of personal self-care activities, i.e. eating, drinking, washing and dressing, environmental issues to improve safety and motor function, cognitive assessment and appropriate intervention.

Speech and language therapy for improvement of vocal loudness and pitch range, including speech therapy programmes i.e. Lee Silverman Voice Treatment (LSVT), teaching strategies to optimise speech intelligibility, ensuring an effective means of communication is maintained throughout PD, including use of assistive technologies , review and mx to support safety and efficiency of swallowing and to minimise risk of aspiration.

Consider Palliative care in PD during all phases. Give opportunity to discuss end of life issues with PD and carers.

At present there is no agent that slows down the progression of PD. The NHS requires neuroprotectants to ↓ the burden of disability caused by PD, thereby ↓ the direct and indirect costs of caring for an ↑ patients with PD.

Which patients with PD and dementia benefit from cholinesterase inhibitors and/or memantine, and is the use of these agents cost effective? A recent systematic review indicated that 24–31% of PD have dementia, and that 3–4% of the dementia in the general pop is due to PD dementia (PDD). The estimated prevalence of PDD in the general pop aged ≥65 is 0.2–0.5%. PDD is a/w ↑mortality, care-giver stress, and nursing home admission.

PROSTATE CANCER (NICE FEB 2008)

Every year there are 34,986 new cases in England and Wales and 10,000 deaths. 20% occur in men < 65 years.

- Healthcare professionals should adequately inform men with prostate CA and their partners or carers about the effects of prostate CA and the tx options on their sexual function, physical appearance, continence and other aspects of masculinity. Support men and their partners or carers in making tx decisions, taking into account the effects on quality of life as well as survival.
- To help men decide whether to have a prostate biopsy, discuss with them their PSA level, DRE findings (including an estimate of prostate size) and comorbidities, together with their risk factors (including age and black African or black Caribbean ethnicity) and any h/o a previous negative prostate biopsy. The serum PSA level alone should not automatically lead to a prostate biopsy.
- Offer active surveillance first to low-risk localised prostate CA who are considered suitable for radical tx.
- Men undergoing radical external beam radiotherapy for localised prostate CA should receive a min. dose of 74 Gy to the prostate at < 2 Gy per fraction.
- Ensure patients and partners have early and ongoing access to specialist erectile dysfunction services.
- Ensure that men with troublesome urinary sxs after tx have access to specialist continence services for assessment, dx and conservative tx. This may include coping strategies, along with pelvic floor muscle re-education, bladder retraining and pharmacotherapy.

- Refer intractable stress incontinence to a specialist surgeon for consideration of an artificial urinary sphincter.
- Biochemical relapse (a rising PSA) alone should not prompt an immediate change in treatment.
- Hormonal treatment is not routinely recommended for men with prostate CA who have a biochemical relapse unless they have: symptomatic local disease progression, or any proven mets, or a PSA doubling time < 3 months.
- When men with prostate CA develop biochemical evidence of hormone-refractory disease, their tx options should be discussed by the urological cancer multidisciplinary team (MDT) with a view to seeking an oncologist and/or specialist palliative care opinion, as appropriate.
- Ensure that palliative care is available when needed and is not limited to the end of life. It should not be restricted to being associated with hospice care.

Table 1 Risk stratification for men with localised prostate cancer.

	PSA		**Gleason score**		**Clinical stage**	**Mx**
Low risk	< 10 ng/ml	**and**	≤ 6	**and**	T1-T2a	Active surveillance
Intermed risk	10–20 ng/ml	**or**	7	**or**	T2b-T2c	Active surveillance
High risk	> 20 ng/ml	**or**	8-10	**or**	T3-T4	Systemic/ Radiotx

Management

- Men with localised prostate CA who have chosen a watchful waiting regimen and who have evidence of significant disease progression (rapidly rising PSA level, bone pain or adverse findings on biopsy) should be reviewed by a member of the urological cancer MDT and offered radical tx.
- Men with low-risk localised prostate CA considered suitable for radical tx should 1st be offered active surveillance. Active surveillance is particularly suitable for a subgroup with low-risk localised prostate CA

who have clinical stage T1c, a Gleason score of 3+3, a PSA density of < 0.15 ng/ml/ml and who have CA in < 50% of their total number of biopsy cores with < 10 mm of any core involved.

- Discuss active surveillance as an option for intermediate-risk localised prostate CA.
- Active surveillance should include at least one re-biopsy and may be performed in accordance with the ProSTART protocol.

Locally advanced prostate CA
- It covers a spectrum of disease from a tumour that has spread through the capsule of the prostate (T3a) to large T4 cancers that may be invading the bladder or rectum or have spread to pelvic LNs.

Systemic treatment
- Neoadjuvant and concurrent LHRHa (agonist) tx is recommended for 3 to 6 months in men receiving radical radiotherapy for locally advanced prostate CA. (Know the side effects of goserelin GnRHa.)
- Adjuvant hormonal treatment in addition to radical prostatectomy is not recommended, even in men with margin-positive disease, other than in the context of a clinical trial.
- Adjuvant hormonal therapy is recommended for a min of 2 years in men receiving radical radiotherapy for locally advanced prostate cancer who have a Gleason score of ≥ 8.
- Bisphosphonates should not be used for the prevention of bone mets in men with prostate cancer.

Radiotherapy
- Clinical oncologists should consider pelvic radiotherapy in men with locally advanced prostate cancer who have a > 15% risk of pelvic lymph node involvement and who are to receive neoadjuvant hormonal therapy and radical radiotherapy.
- Immediate post-op radiotherapy after radical prostatectomy is not routinely recommended, even in men with margin-positive disease, other than in the context of a clinical trial
- High-intensity focused ultrasound and cryotherapy are not recommended for men with locally advanced prostate cancer other than in the context of controlled clinical trials comparing their use with established interventions.

Metastatic prostate cancer

- Hormonal therapy
- Offer bilateral orchidectomy to all with met prostate CAas an alternative to continuous LHRHa treatment.
- Combined androgen blockade is not recommended as a 1st-line treatment for men with metastatic prostate CA.
- For men with met prostate cancer who are willing to accept the adverse impact on overall survival and gynaecomastia in the hope of retaining sexual function, offer anti-androgen monotherapy with bicalutamide (150 mg).
- Begin androgen withdrawal and stop bicalutamide tx in men with met prostate CA who are taking bicalutamide monotherapy and who do not maintain satisfactory sexual function
- Intermittent androgen withdrawal may be offered to men with met prostate CA providing they are informed that there is no long-term evidence of its effectiveness.

Managing the complications of hormonal tx

- Synthetic progestogens (orally or parenterally) are 1st -line treatment for the management of troublesome hot flushes.
- Men starting long-term bicalutamide monotherapy (> 6 months) should receive prophylactic radiotherapy to both breast buds within the 1st month of treatment. A single fraction of 8 Gy using orthovoltage or electron beam radiotherapy is recommended.
- If radiotherapy is unsuccessful in preventing gynaecomastia, consider weekly tamoxifen.
- If starting androgen withdrawal tx advise that regular resistance exercise ↓fatigue and improves quality of life.

Hormone-refractory prostate cancer

- When men with prostate CA develop biochemical evidence of hormone-refractory disease, their tx options should be discussed by urological cancer MDT with a view to seeking an oncologist +/or specialist palliative care opinion.
- Docetaxel is recommended, within its licensed indications, as a tx option for men with hormone-refractory prostate CA only if their Karnofsky performance-status score is ≥ 60%.

QUANTITATIVE AND QUALITATIVE STUDIES

Quantitative Studies excels at summarizing large amounts of detailed numerical data and reaching generalizations based on statistical projections, to provide statistical info about research questions so accurate conclusions can be drawn from the data. Methodologies: cohort, case control, RCTs.

Qualitative Studies TRIANGULATION: Interview, focus groups and observation. Generalizing across a sample of interviews or written documents. Excels at "telling the story" from the pt's viewpoint, providing the rich descriptive detail that sets quantitative results into their human context. **Pros**: enables you to describe the phenomena of interest in great detail, in the original language of the research participants. Identifying informants who directly experienced the phenomenon in question, interviewing, and then editing the interviews so that they address the question of interest. **Cons:** hard to determine what the generalizable themes may be with so much detail. **Generally exploratory in nature, involving small sample sizes.** Unlike quantitative research, questions are more loosely structured to promote discussion. Understand the reasoning behind someone's decision to use (or not use) a product or service. **Used to explore-or even to generate-new ideas**. Asks probing questions to both identify the factors that drive user behaviour and to understand why the factors are important. A qualitative approach: 1) Get feedback 2) Ask customers about the features and benefits they want, and 3) Find out if and why adding this functionality to the product will impact their decision to use it. Success stories, user needs analysis. Qualitative methodologies-1-on-1 interviews, phone-or internet-based focus groups. **Mixed Methodologies** - Qualitative research to generate initial ideas and insights into research questions. Because the sample size is small, and the survey questions are exploratory in nature, the results of a qualitative research study are often "directional" rather than conclusive. **Initial ideas can be a pre-cursor to a FU quantitative study**. **Formally testing these ideas with a large sample**, and applying the appropriate sampling plan, will provide statistical validation of the info needed to make important decisions.

(nGMS) NHS National QOF 2007/8, published Oct 2008

Clinical Domain: 80 indicators across 19 clinical areas: CHD, ht F, Stroke and TIA, HTN, DM, COPD, Epilepsy, Hypothyroidism, CA, Palliative Care, MH, Asthma, Dementia, Depression, CKD, AF, Obesity, Learning disabilities, Smoking. **Organisational Domain**: 42 indicators across 5 areas: records + info; patient info; education and training; practice mx; medicines mx. Patient

Experience Domain: 4 indicators relating to length of consultations and patient surveys. **Additional Services Domain**: 8 indicators across 4 service areas: cervical screening, child health surveillance, maternity services and contraceptive services. Holistic care payment.
15-75 yo - record smoking status (give anti-smoking advice/ support). ≥ 45 yo record BP at least 5 yrly.

IHD On CHD register for angina/ MI and LVF register for LVF. Angina - refer to cardiology OR had exercise ECG OR referred to rapid access chest pain clinic. Refer all new anginas for specialist assessment and all new LV dysfunction for echos. Record annually smoking status (unless never smoked) and annual advice to stop. BP and cholesterol annually. Annual flu jab (last Sept-March). Get BP ≤150/90 or add 'on max BP rx' , TC ≤ 5 mmol/l (or exception report annually if on max tolerated rx). Unless C/I, put all patients on ASA, B-blocker, ACEI/ A2 blockers (if MI after April 1, 03, or LVF). On aspirin, clopidogrel or warfarin?

Stroke/ TIA On stroke/ TIA register (haemorrhagic CVA, ischaemic CVA (CVA clot), CVA type unspecified (no scan done), TIA (SAH is excluded). Refer all new stroke patients for CT/ MRI. Had CT or MRI 3/12 before to 1 year after? Record annual smoking status (unless never smoked) and annual stop smoking advice. BP and TC annually. Get BP ≤ 150/90; TC ≤ 5. Annual flu jab. Give ASA or alternate anticoagulant (dipyridamole/ clopidogrel/ warfarin) - unless C/I or S/E recorded

Hypertension Record smoking status at least once and stop smoking advice. BP q 6-9 months (1 July onward for each yr). Get BP ≤ 150/90 or exception report annually if on max tolerated rx.

Diabetes mellitus On diabetes register if > 16 yrs. Record annual BMI, smoking status and advice to stop. Flu jab. Annually HbA1C and BP. % BP at diabetic target (145/85) or add 'max tolerated BP rx' annually. Record annual retinal screening (advised to attend for retinal screen OR under ophthalmologist OR under retinal screener), peripheral pulses (in R/L leg), neuropathy testing (OR under care of DM foot screen), microalbuminuria testing (OR persistent proteinuria is present - needs 24h urine. One entry ever is OK), creat and TC ≤ 5 mmol/l (or add 'max tolerated lipid rx') Get HbA1C ≤ 10% in last yr) or ≤ 7.4. Treat proteinuria/microalbuminuria with ACEI.

Hypothyroidism - On register. Acquired/ congenital. Annual TFTs.

Cancer - Review patients within 6/12 of dx to assess support needs and 2° care

Asthma code with rx in last 2/12. PEF checked ever (age 8+) - from 3/12 before dx entered. Age 14-19, age 20 +annual smoking status with annual stop smoking advice. Annual asthma review (PEF and inhaler technique check). Age 16+, annual flu jab.

COPD On COPD register. Spirometry once for dx made after 1/4/03 or once ever on all patients (from 3/12 before to 1 yr after dx). FEV1 repeated every 2 years. Smoking health ed in smokers annually. Flu jab. Inhaler technique shown/ observed every 2 years.

Mental Health Longterm MH problems + those ever rx lithium who agree to F/U. Exclude those who request it. Annual mental health R/V. Lithium checked in last 6/12 if on lithium or MH med review. TFTs in last yr (if on lithium). Lithium in normal range in last 6/12 (or if lithium is in normal range for this particular patient)

Epilepsy If > 16 yrs and on rx - fit freq record in last year or add 'max tolerated epileptic rx' annually; treatment R/V in last yr; seizure-free in last 12 months?

QOF Exception Reporting: allow practices to pursue QOF and not be penalized, i.e. patients do not attend for review, or where a rx cannot be prescribed due to a C/I or S/E. Criteria: patients recorded as refusing to attend review wo have been invited at least 3 times during the preceding 12 mos. Patients it is not appropriate to review the chronic disease parameters due to particular circumstances (terminal illness, extreme frailty), patients newly diagnosed w/n the practice or who have recently registered who should have measurements made w/n 3 months and delivery of clinical standards w/n 9 months (BP, cholesterol measurements w/n target levels); patients who are on max tolerated doses of meds whose levels remain sub-optimal; patients for whom prescribing a med is not clinical appropriate (allergy, another C/I or adverse reaction); where a pt has not tolerated med; where a pt does not agree to ix or tx (informed dissent) and this has been recorded in records; where a pt has a supervening condition which makes tx of their condition inappropriate (cholesterol reduction where the patient has liver disease); where an ix service or 2° care service is unavailable.

SCHIZOPHRENIA (NICE Dec 2002)
Advance directives regarding choice of treatment recommended.
Types: paranoid, catatonic, residual, undifferentiated, disorganized.

Positive symptoms: an excessive or distorted version of normal functions. Includes **Schneider's first-rank symptoms** (set of symptoms designated by Kurt Schneider (1959) as the most important diagnostic indicators of schizophrenia: delusions, hallucinations, thought insertion or removal, thought broadcasting) as well as disorganized thought processes (speech) and disorganized or catatonic behaviour. Disorganized thought processes are marked by looseness of associations (rambles from topic to topic in a disconnected way); tangentially, (gives unrelated answers to questions); and "word salad" (patient's speech is so incoherent that it makes no grammatical or linguistic sense). Disorganized behaviour: difficulty with any type of purposeful or goal-oriented behaviour (personal self-care or preparing meals), dressing in odd or inappropriate way, public sexual self-stimulation, or agitated shouting or cursing.

Negative sxs: because they represent the lack or absence of behaviours. The 3 negative sxs considered diagnostic are a lack of emotional response (affective flattening), poverty of speech, and absence of volition or will.

Initiation of treatment at the 1st episode
- Refer urgently to 2° MH for assessment & development of care plan
- Early intervention (specialist, pharmacological, psychological, social, occupational and educational).
- Consider referral to crisis resolution and home treatment teams, acute day hospitals or inpatient.
- GP to consider starting atypical antipsychotic drugs when acute sxs of schizophrenia, before seen by a psychiatrist if necessary, following discussion with psychiatrist and refer urgently.

Acute phase
- Oral atypical antipsychotic (amisulpiride, olanzapine, quetiapine, risperidone, zotepine) 1st line (less EPS).
- CMHTs, assertive outreach teams, crisis resolution and home tx teams (augment services provided by 1st two), acute day hospital
- May need rapid tranquilisers (IM lorazepam, haloperidol or olanzapine). Resuscuscitation, flumazenil.

- Conventional antipsychotic med - 300-1000 mg chlorpramazine/ day for 6/52. Skin photosensitivity.
- Consider psychoanalytic/ psychodynamic principles. CBT should be available. Family interventions.
- High risk of relapse consider continue antipsychotic drugs for up to 1-2 years after a relapse and monitor for at least 2 yrs after the last acute episode after withdrawal from medication.

Promoting recovery
Practice case registers. Refer if treatment adherence is a problem, poor response to treatment, co-morbid substance misuse, risk to self/ others ↑, when a person with schizophrenia first joins a GP practice, refer for assessment and care programme.
Relapse prevention: depot antipsychotics. Atypicals for tx-resistant schizophrenia (6-8/52 of rx). Use trial of clozapine, olanzapine or risperidone.

STATISTICS

Sick and Fit

			Disease	
			+	-
Exposure/Test		+	A	B
		-	C	D

Sensitivity = proportion of TP correctly identified by test = A/ (A+C) (SICK)
Specificity= proportion of TN correctly identified by test = D/ (B+D) (FIT)
Positive predictive value = proportion of patients with positive test correctly identified as diseased = A/ (A+B)
Negative predictive value = proportion of patients with negative test correctly identified as disease-free = D/(C+D)

Example 1 100 population, prevalence of disease is 1% = 1 + disease, 99 °dis. Test has 100% sensitivity, 95% specificity:

		Disease		
		+	-	
Test	+	1	5	6
	-	0	94	94
		1	99	100
		TP	TN	tot pop

PPV = 1/6 = 17%   NPV = 94/94= 100%

Example 2 Population 500 in a small town in China. 100 inhabitants have AIDS. 80/100 with the disease, test positive correctly. 50 test + who have no disease.

			Disease +	-	
Test	+		80	50	= 130
	-		20	350	= 370
100	400	500			
			TP	TN	

Prevalence of disease is 100/500 = 20%
Sensitivity of test = 80/100 = 80%   Specificity of test = 350/400 = 87.5%
PPV = 80/130 = 62%   NPV = 350/370 = 95%

Example 3 Mammo screening test. Specificity 80%. Sensitivity 90%. Population of 1000 females aged 50-65 with prevalence of breast CA 10%. PPV = 90/270 = 33% NPV = 720/730 = 99%

COHORT STUDIES

RELATIVE RISK - disease rate in exposed group divided by disease rate in non exposed group;
incidence of disease in exposed population divided by incidence of disease in non-exposed (NE) population

DIVISION

$$\frac{A}{A+B} \div \frac{C}{C+D} = \frac{\text{dis in exp}}{\text{dis in exp} + {}^\circ\text{dis in exp}} \div \frac{+\text{dis in NE}}{+\text{dis in NE} + {}^\circ\text{dis in NE}}$$

i.e. relative risk = 1 = exposure is not associated with ↑ or ↓ risk of disease; i.e. no effect on the disease

ATTRIBUTABLE RISK - excess risk (or outcome) attributable to a given exposure

SUBTRACT

$$\frac{A}{A+B} - \frac{C}{C+D} = \frac{\text{dis in exp}}{\text{dis in exp} + {}^\circ\text{dis in exp}} - \frac{+\text{dis in NE}}{+\text{dis in NE} + {}^\circ\text{dis in NE}}$$

CASE-CONTROL STUDIES

Exposure rate for cases
$$\frac{A}{A+C} = \frac{\text{disease in exposed}}{\text{dis in exposed + dis in NE}}$$

Exposure rate for controls
$$\frac{B}{B+D} = \frac{°\text{disease in exposed}}{°\text{dis in exposed} + °\text{dis in NE}}$$

Odds Ratio is the estimate of relative risk:
$$\frac{A \times (C+D)}{C \times (A + B)} = \frac{\text{dis in exp}}{\text{dis in NE}} \times \frac{(\text{dis in NE} + °\text{dis in NE})}{(\text{dis in exp} + °\text{dis in exp})} \sim \frac{A \times D}{C \times B}$$

Example 4   DVT

		+	-	
Pill Use	+	A=100	B=200	100/300 ♀ with DVT (dis) use the pill
	-	C=200	D=800	200/1000 controls without DVT are on pill
		300	1000	

Odds Ratio = $\frac{A \times D}{C \times B} = \frac{100 \times 800}{200 \times 200} = 2$ = Pill users have a 2 x > risk of DVT.

COHORT:

Example 5   100 in 1000 F smokers at risk of cervical CA. 60 in 1000 F nonsmokers at risk of cervical CA.

		Disease (Cervical CA)		
		+	-	
Exposed to tob	+	100	900	1000
	-	60	940	1000

1. Absolute risk (incidence) in general ♀ pop = 60/1000= 6 in 100= 1 in 17

2. Relative risk (incid of dis in Exp pop ÷ incid dis NE) = $\frac{100}{1000} \div \frac{60}{1000}$
= 1.67

3. Attributable risk (incid in Exp pop - incid dis NE) = $\frac{100}{1000} - \frac{60}{1000} = \frac{40}{1000}$
= 0.04

Example 6   Smoking ↑ risk of lung CA by 40%. Smoking ↓ risk of PD by 20%.

1. Relative risk of lung CA in smokers vs nonsmokers? $1 + (40\%$ of $1) = 1.4$
2. Relative risk of PD in smokers vs nonsmokers? $1 - (20\%$ of $1) = 0.8$

NNT (numbers needed to treat)

Definition # of individuals who need to be treated to achieve one desired outcome

- reciprocal of the reduction in absolute risk (absolute risk reduction or ARR); **1/ARR%**
- depends on prevalence of given disease, i.e. high AR or incidence of heart disease have low NNT to prevent 1 death

ARR = proportion of treated group with desired outcome - population of controls with desired outcome

ARR% = % treated group with desired outcome - % controls with desired outcome

Example 7

% giving up smoking in treated group with bupropion = $10/100$ $= 10\%$
% giving up smoking in control group   = $12/200$ $= 6\%$
ARR% $= 10\% - 6\% = 4\%$
NNT $= 1/ARR\% = 1/4\% = 25$ people needed to be given bupropion to cause 1 person to quit
Number of person-years of observation: 10 p-y $= 1$ person observed for 10 yrs OR 5 patients for 2 yrs.

METHODS (STUDY DESIGNS)

CROSS-SECTIONAL SURVEYS

- **prevalence** (proportion)studies, descriptive study which provides a **snapshot** of the pop in question
- the +/- of disease can be assessed and compared with age, sex or body weight
- **cannot assess incidence of exact timing of exposure or cause or effect**
- ex use of Mirena by doctors – current use % vs. age group; random postal questionnaire between 1990 and 1995

CASE-CONTROL STUDIES

- case (subjects with disease) and controls (same population as cases but no disease)

- **retrospective** studies look for **exposure** or presence of factor **association not causation**
- assess by questionnaire, interviews or medical records
- investigate hypothesis and may be followed up by cohort or intervention studies
- **cannot calculate incidence (new cases) of disease, true relative risk, attributable risk**
- **can calculate odds ratio** (estimate of relative risk) if incidence of disease in general population < 5%, control group representative of general population, and cases + controls free from selection bias
- pros (**small number of subjects**, assess **rare conditions**, quick + cheap, multiple exposures)
- cons (**bias**, retrospective (recall-bias), hard to establish causal relationship and to select controls
- ex lung CA in shipyard workers – cases vs. controls (relatives)

COHORT STUDIES
- **prospective** (looking ahead), for **rare exposures and several outcomes** may be studied
- pros **(can calculate incidence directly, relative risk, attributable risk);** selection bias is less likely
- several exposures can be evaluated from exposure to the same agent
- cons (hard to interpret causal relationships, expensive as large pop studied over long time, loss of FU can affect validity)
- ex (study starts 1980s and ends 20 yrs later, series of observations made on subjects)

HISTORIC COHORT STUDY (retrospective cohort study) A research study in which the medical records of groups of individuals who are alike in many ways but differ by a certain characteristic (F nurses who smoke and those who do not smoke) are compared for a particular outcome (lung CA).

CROSSOVER STUDY is a clinical trial in which patients are given all of meds to be studied, or 1 med + placebo in random order. Done on patients with chronic disease to control their sxs. Pros: over a simple double-blind study that the variability between patients is minimized because each pt crossing over in effect serves as own control. Cons: cannot track long term effects and cannot test curative txs after one another or before a placebo. The order in which txs are administered may affect the outcome, i.e. the order may have adverse effects on each other. In practice carry-over between txs is dealt with by the use of a wash-out period between txs, or by making observations sufficiently later after the start of a tx period to ↓ carry-over effects.

GROUNDED THEORY is a qualitative research method in which the theory is generated from the data, rather than vice versa. Popularized by Glaser and Strauss, the researcher collects data about a single subject without any preconceived idea concerning its content or structure. The theory is 'grounded' in the data rather than the data having been collected to test a pre-existing theory, a procedure which is likely to bias the data used.

OBSERVATIONAL STUDY –for experiments prone to selection bias.

RCTs (CLINICAL)
- intervention studies prompted by findings of case-control, cohort or cross-sectional survey
- intervention (treated) group vs. control group allocated randomly
- **gives best evidence of cause and effect;** i.e. tx with interferon for MS

CONCLUSIONS - difference between general population and sample study group due to chance sampling variation, study biases, confounding factors, or true difference

NULL HYPOTHESIS - there is no difference between the 2 groups; the intervention has no effect

TEST OF STATISTICAL SIGNIFICANCE - based on null hypothesis ex. t-test, chi-squared

ρ VALUE - probability of this result occurring by chance, if null hypothesis is true
- ρ value < 0.05 indicates statistical significance, < 1/20 due to chance
- ρ value irrelevant if study + biases, confounding factors, small sample size

CONFIDENCE INTERVAL - % probability that true value lies within the confidence limits
- ex 95% CI = 95% chance true value lies w/n stated limits or 5% chance lies outside
- Desire narrow CI less chance of variability in sample and greater certainty
- Suppose an opinion poll predicted that, if the election were held today, the Conservative party would win 60% of the vote. The pollster may attach a 95% confidence level to the interval 60% plus or minus 3%. He strongly believes that the Conservative party will get between 57% and 63% of the total vote.

RELATIVE RISK - Results of cohort studies; ratio of disease incidence in exposed vs. NE
- RR = 1 (no association between exposure + outcome) Ex RR for cholelithiasis with obesity = 2(0.8-2.7) 95% CI

STANDARDISED MORTALITY RATIO number of observed deaths ÷ by number of expected deaths.

CONFOUNDING FACTORS
- spurious association between 2 factors due to 3rd factor (age, race, sex, socioeconomic group, tobacco)
- positive confounding (smoking vs. risk of heart disease - lower socioeconomic are unhealthy)
- negative confounding (neutralizes degree of association)

N-of-1 TRIALS randomised, double-blinded, multiple crossover comparisons of an active drug vs. a placebo or an alternative tx in a single pt. Objectivity of this trial in COAD patients is superior to standard practice. Top of hierarchy evidence. Useful for prolonged chronic conditions in which the proposed tx has a rapid onset of action with readily observable effects and ceases to act after it is discontinued. RCTs are average estimates of benefit and harm and patients with multiple chronic disease might attenuate benefit.
Cons: substantial effort, lack of intellectual and admin experience, funding.

CHI-SQUARE is a non-parametric test of statistical significance for bivariate tabular analysis.
- Tests whether or not 2 samples are different enough in a particular aspect so that we can generalize that the populations from which the samples were drawn are also different. It is a rough estimate of confidence.
- Is there a relationship between any 2 variables in the data? How strong is this relationship?
- The sample must be randomly drawn from the population.
- Data must be reported in raw frequencies at not %.
- Independent measured variables. Values on variables must be mutually exclusive and exhaustive.
- Observed frequencies can't be too small.

STUDENT'S t TEST tests whether the means of 2 groups (i.e. control and treatment group) are statistically different.
- T-test is positive if the 1st mean is larger than the 2nd and negative if it is smaller.
- Alpha (risk) level should be set at 0.05 (5x out of 100 there is a stat. significant difference between the means).
- Degrees of freedom (df) for the test is the sum of the persons in both groups minus 2.
- If the t-value is large enough to be significant (computer program or t-value chart), it means that the difference between the means for the 2 groups is different.

STROKE (NICE – May 2008)

Rapid recognition of symptoms and diagnosis

Use sudden onset of neurological symptoms a validated tool, FAST (Face Arm Speech Test), outside hospital to screen for a dx of stroke or TIA. In patients with sudden onset of neurological symptoms, exclude hypoglycaemia. Patients who are admitted to A&E with a suspected stroke or TIA should have the dx established rapidly using a validated tool, i.e. ROSIER (Recognition of Stroke in the Emergency Room).

Patients who have had a suspected TIA who are at high risk of stroke (ABCD^2 score of $\geq$ 4) should have:

aspirin (300 mg daily) started immediately

specialist assessment and ix w/n 24 hrs of onset of sxs

measures for 2° prevention introduced as soon as the dx is confirmed, discuss individual risk factors.

Treat crescendo TIA ($\geq$ 2 TIAs/wk) as high risk of stroke, even though they may have an ABCD^2 score of $\leq$ 3.

Some who have had a stroke or TIA have narrowing of the carotid artery that may require surgical intervention. Carotid imaging is required to define the extent of carotid artery narrowing. The use of carotid stenting was also reviewed by the GDG. However, no evidence for early carotid stenting was found on which the GDG felt they could base a recommendation. Patients who have had a suspected TIA (sxs and signs have completely resolved < 24 hrs) should be assessed by a specialist (<1 week of sx onset) before a decision on brain imaging is made.

Patients who have had a suspected TIA who are at high risk of stroke (an ABCD^2 score of $\geq$ 4, or with crescendo TIA) in whom the vascular territory or pathology is uncertain should undergo urgent brain imaging (preferably diffusion-weighted MRI). All with suspected non-disabling stroke or TIA who after specialist assessment are considered as candidates for carotid endarterectomy should have carotid imaging < 1 wk of onset of sxs. Patients who present > 1 wk after their last sx of TIA has resolved should be managed using the lower-risk pathway.

Refer for urgent carotid endarterectomy and carotid stenting for stable neurological sxs from acute non-disabling stroke or TIA who have symptomatic carotid stenosis of 50–99% according to the NASCET (North American Symptomatic Carotid Endarterectomy Trial) criteria, or 70–99% according to the ECST (European Carotid Surgery Trialists' Collaborative

Group) criteria within 1 wk of onset of stroke or TIA sxs; undergo surgery within a max of 2 wks of onset of stroke or TIA sxs; receive best medical treatment (BP control, antiplatelet agents, cholesterol lowering via diet and drugs, lifestyle advice). Carotid imaging reports should clearly state which criteria (ECST or NASCET) were used when measuring the extent of carotid stenosis.

Specialist care for people with acute stroke All suspected stroke should be admitted directly to a specialist acute stroke unit following initial assessment, either from the community or from the A&E. Brain imaging should be performed immediately if any of the following apply:
- indications for thrombolysis or early anticoagulation tx
- on anticoagulant tx
- a known bleeding tendency
- a depressed level of consciousness (Glasgow Coma Score < 13)
- unexplained progressive or fluctuating sxs
- papilloedema, neck stiffness or fever
- severe headache at onset of stroke sxs.

Thrombolysis with alteplase is recommended for the tx of acute ischaemic stroke when used by doctors trained in the mx of acute stroke and administered only within a well organised stroke service with: staff trained in delivering thrombolysis and in monitoring for any complications a/w thrombolysis, level 1 and level 2 nursing care staff trained in acute stroke and thrombolysis, immediate access to imaging and re-imaging, and staff trained to interpret the images. Staff in A&E, if appropriately trained and supported, can administer alteplase for the treatment of acute ischaemic stroke provided that patients can be managed within an acute stroke service with appropriate neuroradiological and stroke physician support. Protocols should be in place for the delivery and mx of thrombolysis, including post-thrombolysis complications.

Aspirin and anticoagulant tx - People with acute ischaemic stroke
All who have had a dx of 1° intracerebral haemorrhage excluded by MRI should be given, ASAP or < 24 hours:
- aspirin 300 mg orally if they are not dysphagic or
- aspirin 300 mg rectally or by enteral tube if they are dysphagic.
- Thereafter, continue aspirin 300 mg until 2 weeks after the onset of stroke symptoms, at which time initiate definitive long-term antithrombotic treatment. People being discharged < 2 weeks can be started on long-term treatment

earlier. Give PPI if previous dyspepsia a/w aspirin is reported. Give an alternative antiplatelet agent if allergic to or genuinely intolerant of aspirin. Anticoagulation tx should not be used routinely for the tx of acute stroke.

Patients with acute venous stroke diagnosed with cerebral venous sinus thrombosis (including those with 2° cerebral haemorrhage) should be given full-dose anticoagulation treatment (initially full-dose heparin and then warfarin [INR 2–3]) unless there are comorbidities that preclude its use. Stroke 2° to acute arterial dissection should be treated with either anticoagulants or antiplatelet agents, preferably as part of a RCT to compare the effects of the 2 treatments. Acute ischaemic stroke associated with antiphospholipid syndrome should be managed in same way as patients with acute ischaemic stroke without antiphospholipid syndrome.

Reversal of anticoagulation treatment in patients with haemorrhagic stroke
Clotting levels with a 1° intracerebral haemorrhage who were receiving anticoagulation treatment before their stroke (and have elevated INR) should be returned to normal ASAP, by reversing the effects of the anticoagulation treatment using a combination of prothrombin complex concentrate and IV vitamin K.

Anticoagulation treatment for other comorbidities
Treat disabling ischaemic stroke who are in atrial fib with aspirin 300 mg for the 1st 2 wks before anticoagulation.
In patients with prosthetic valves who have disabling cerebral infarction and who are at significant risk of haemorrhagic transformation, anticoagulation treatment should be stopped for 1 week and aspirin 300 mg substituted.
Ischaemic stroke and symptomatic proximal DVT or PE should receive anticoagulation treatment in preference to treatment with aspirin unless there are other C/Is to anticoagulation. Haemorrhagic stroke and symptomatic DVT or PE should have treatment to prevent the development of further PE using either anticoagulation or a caval filter.
Statin Immediate initiation of statins is NOT recommended in acute stroke. Continue statin treatment if already on.

Maintenance or restoration of homeostasis
A key element of care for acute stroke is the maintenance of cerebral blood flow and oxygenation to prevent further brain damage after stroke. **Supplemental O_2 tx** receive only if their O_2 sat drops < 95%. **Blood sugar**

control Acute stroke should be treated to maintain a blood glucose concentration between 4 and 11 mmol/litre.

Optimal insulin tx, which can be achieved by the use of IV insulin and glucose, should be provided to all diabetics who have threatened or actual MI or stroke. Critical care and emergency departments should have a protocol.

BP control Anti-hypertensive tx with acute stroke is recommended only if there is a hypertensive emergency with ≥ 1 of the following serious concomitant medical issues: hypertensive encephalopathy; hypertensive nephropathy; hypertensive cardiac failure/MI; aortic dissection; pre-eclampsia/eclampsia; intracerebral haemorrhage with SBP > 200 mmHg. Consider BP reduction to ≤ 185/110 mmHg in patients for thrombolysis.

Nutrition and hydration

On admission, screen swallowing before being given any oral food, fluid or meds. If the admission screen indicates problems with swallowing, arrange a specialist assessment of swallowing, preferably w/n 24 hrs of admission and not > 72 hrs afterwards. Suspected aspiration on specialist assessment, or who require tube feeding or dietary modification for 3 days, should be: re-assessed and considered for instrumental exam; referred for dietary advice. Acute stroke who are unable to take adequate nutrition and fluids orally should:

- receive tube feeding with a NG tube w/n 24 hrs of admission
- be considered for a nasal bridle tube or gastrostomy if they are unable to tolerate a NG tube
- be referred to an appropriately trained professional for detailed nutritional assessment, advice and monitoring.

Oral nutritional supplementation

All hospital inpatients on admission should be screened for malnutrition and its risk. Repeat screening weekly for inpatients. Screening should assess BMI and % unintentional wt loss and should also consider the time over which nutrient intake has been unintentionally reduced and/or the likelihood of future impaired nutrient intake. The Malnutrition Universal Screening Tool (MUST) may be used to do this.

When screening for malnutrition and the risk of malnutrition, be aware that dysphagia, poor oral health and reduced ability to self-feed will affect nutrition. Routine nutritional supplementation is not recommended for people with acute stroke who are adequately nourished on admission. Nutrition support should be initiated for those who are at risk of malnutrition - oral nutritional supplements, specialist diet advice +/or tube feeding. All assess hydration on admission, review regularly and manage so that normal hydration is maintained.

Early mobilisation and optimum positioning of people with acute stroke a key element of acute stroke care. Sitting up ASAP will help to maintain O_2 sats and ↓ the likelihood of hypostatic pneumonia. Mobilise ASAP (when their clinical condition permits) as part of an active mx programme in a specialist stroke unit.

Avoid aspiration pneumonia, a complication of stroke that is associated with ↑ mortality and poor outcomes. Give food and fluids in a form that can be swallowed without aspiration, following specialist assessment of swallowing.

Surgery with acute stroke with intracerebral haemorrhage or severe middle cerebral artery infarction.

Evidence that neurosurgical tx may be indicated for a very small number. Surgical referral for acute intracerebral haemorrhage - Stroke services should agree protocols for the monitoring, referral and transfer to regional neurosurgical centres for the mx of symptomatic hydrocephalus.

Patients with intracranial haemorrhage should be monitored by specialists in neurosurgical or stroke care for deterioration in function and referred immediately for brain imaging when necessary.

Previously fit patients should be considered for surgery following 1° intracranial haemorrhage if they have hydrocephalus.

Patients with any of the following rarely require surgical intervention and should receive medical tx initially:

small deep haemorrhages; lobar haemorrhage without either hydrocephalus or rapid neurological deterioration

a large haemorrhage and significant comorbidities before the stroke; a score on the GCS of < 8 unless this is because of hydrocephalus; posterior fossa haemorrhage.

Surgical referral for decompressive hemicraniectomy for middle cerebral artery infarction who meet all of the criteria below < 24 hours of onset of sxs and treated within a max of 48 hrs: Aged ≤ 60 yrs. Clinical deficits suggestive of infarction in the territory of the MCA, with a score on the National Institutes of Health Stroke Scale (NIHSS) of > 15. ↓ in the LOC to give a score of ≥ 1 on item 1a of the NIHSS. Signs on CT of an infarct of at least 50% of the MCA territory, with or without additional infarction in the territory of the anterior or posterior cerebral artery on the same side, or infarct volume > 145 cm^3 as shown on diffusion-weighted MRI.

URINARY INCONTINENCE

Hx	length of time, frequency, amount of leakage, fluid intake, present mx and impact on lifestyle
Exam	abdomen, PR, PV, relevant neurological exam
Ix	frequency/ volume chart, residual urine estimate (scan or in/out catheterisation), urine dip/ MSU

Stress incontinence

Symptoms:	leaking with coughing, exercise or laughing
Aetiology:	urethral sphincter incompetence; pelvic floor weakness
Tx:	pelvic floor exercises; urethral appliances; surgery
Referral:	continence specialist nurse; physiotherapist; uro-gynaecologist; urologist

Detrusor overactivity

Sxs:	urgency, frequency (>7/24h), urge incontinence
Aetiology:	idiopathic; 2^0 to neurological disease (multiple sclerosis); atrophic urethritis/ vaginitis; bladder calculus refer to surgeons; cystitis, UTI treat with antibiotics

Tx for idiopathic/ neurological: check residual volume, fluid intake advice, bladder retraining programme, anticholinergics

Tx for atrophic urethritis/ vaginitis: topical oestrogen replacement or systemic HRT

Bladder outlet obstruction

Sxs	voiding inefficiency, continual dribbling, weak flow, hesitancy; incomplete voiding, intermittent stream, straining to void
Aetiology	prostate hypertrophy, urethral stricture, faecal impaction
Tx	refer to prostate assessment clinic, urologist

Detrusor failure

Sxs	voiding inefficiency (see above)
Aetiology	2^0 to neurological disease
Tx	clean intermittent catheterisation if post-micturition residual > 150 ml
Referral	continence specialist nurse, specialist continence service
Confounding factors	α-adrenergic blockers, anticholinergics, Ca channel blockers, diuretics, sedatives

UROLOGY (MALE FERTILITY/ HAEMATURIA)

Semen Analysis: WHO normal averages: volume 1-5 ml, Density 20 m/ml, motility > 60%, morphology > 60% normal
Refer: < 1m/ejaculate, motility< 20%, progression 2/4 (twitch vs. dash), abnormal forms > 85%, congenital lack of vas
deferens for IVF/ ICSI
Male hormone ix: FSH (if 2x↑ = intrinsic problem with testicle), LH, PRL, testosterone, (inhibin B). If normal, then
obstructive cause. Size of testicle should be 5-6cm = % sperm.
Mx of Oligospermia (< 20 m/ml) - varicocoele ligation (NICE claims does not improve pregnancy rate but some of
NICE studies were flawed), antisperm Ab (obsolete, just refer to IVF clinic), IUI, ICSI
Causes of Azoospermia (0m/ml) - obstruction (epididymal (#1), ejaculatory duct, vasal aplasia, vasal, intratesticular)
VS testicular failure (hypergonadotrophism, hypogonadism)
ICSI in obstructive azoospermia = 47% pregnancy success rate. Both partners require operations.
Non-obstructive azoospermia (sperm retrieval - TESA, TESE, microdissection TESSE, multiple biopsies of testicle)

Ix of Asymptomatic Haematuria in the Adult (SIGN 1997)
Detection of microscopic haematuria (≥ 5rbc/hpf or +ve dipstick)

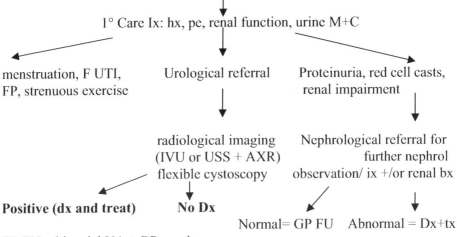

1° Care Ix: hx, pe, renal function, urine M+C

menstruation, F UTI,
FP, strenuous exercise

Urological referral

Proteinuria, red cell casts,
renal impairment

radiological imaging
(IVU or USS + AXR)
flexible cystoscopy

Nephrological referral for
further nephrol
observation/ ix +/or renal bx

Positive (dx and treat)   **No Dx**

Normal= GP FU   Abnormal = Dx+tx

GP FU = biennial UA + BP monitor

VACCINES

HiB	Asplenia, give 2 wks before elective splenectomy. Adults and kids > 1 yo give 2 separate doses.
Influenza	COAD, CHD, chronic liver disease, CRF, DM, IS (prolonged corticosteroid), HIV, asplenia, > 65 yo
Men C	asplenia, kids < 1 yo, unimmunised adults + kids > 1 yo need 2 doses of men C at 2/12 interval
Men ACW135Y	Certain countries in Africa and Hajj/ Umrah pilgrimages to Saudi
Pneumococcal	Age > 65, asplenia or splenic dysfunction from coeliac, COAD, CHD, CRF, chronic liver disease, DM on insulin or oral, I/S, cochlear implant, CSF leak, < 5yo with h/o invasive pneumococcal disease

VAGINAL DISCHARGE

Bacterial vaginosis (Gardnerella vaginalis)	gram-negative rod; fishy odour, bubble-bath, douche, stringy grey-yellow raw egg-white discharge, not STI, salt and pepper (microsopy); rx flagyl 400 mg bd 5/7
Chlamydia trachomatis	obligate intracellular parasite; STI, may be asymptomatic, mucopurulent cervicitis, contact bleeding; endocervical swab; infertility, ectopic, PID; doxycycline 100 mg bd x1/52 or azithromycin1g stat
Gonorrhoea	gram-negative diplococcus; STD, purulent yellow-green discharge, HVS, urethral, rectal swabs; rx ciprofloxacin 500 mg STAT; reculture 3-7d after tx; contact tracing
PID	rx cipro STAT + doxycycline 100 mg bd x 2/52 + metronidazole 400 mg bd x 5/7
Thrush (candidiasis)	white cheesy discharge, pruritis vulvae, superficial dyspareunia; RFs: Addison's or Cushing's, broad-spectrum antibiotics, DM, steroids, IS, pregnancy); rx clotrimazole pessary and cream
Trichomonas vaginalis	STD, mucopurulent, yellow, frothy discharge; strawberry-red cervix; flagellate trophozoite protozoa; rx metronidazole 400 mg bd 5/7

VIAGRA (Sildafenil) (BNF)

Available on NHS rx for: diabetes, MS, PD, polio, prostate CA, severe pelvic injury, single gene neurological disease, spina bifida or spinal cord injury; receiving dialysis for renal failure; have had radical pelvic op, prostatectomy (also TURP), or kidney transplant; were receiving viagra on NHS rx on Sept 14 1998; suffering severe distress (rx by specialist centre). Endorse rx SL. **C/I**: ↓BP if on nicorandil, nitrates, amlodipine; ↑ plasma concentration viagra, ↑ s/e (antivirals, grapefruit juice, cimetidine); ↓plasma concentration sildafenil (bosentan antihypertensive).

2096873R00129

Printed in Great Britain
by Amazon.co.uk, Ltd.,
Marston Gate.